Our
PFAS
Legacy

A Ticking Time Bomb We Can't Escape

Douglas B Sims, PhD

i

DB Sims, PhD

For more information, or to book an event, contact:
dsims@simsassociates.net

Book design by DB Sims
Cover design by DB Sims
Cover picture purchased from iStock (*Murat Deniz*)

ISBN – Paperback: 978-1-966739-06-7
ISBN – eBook: 978-1-966739-07-4

First Edition: April 2025

Please leave a review on Amazon, Goodreads, or any site that you purchased this book as a review is part of the overall experience.

DB Sims, PhD

Table of Contents

Acknowledgments

I am profoundly grateful to my wife, whose steadfast support, wisdom, and love have been my anchor and inspiration. The journey we've shared over the past 34 years has enriched every chapter of my life and this project. Your encouragement has lifted me through challenges, and your insights have shaped my perspective in ways reflected on every page.

To our children, thank you for filling our lives with joy and growth, teaching us both the rewards and trials of parenthood. Watching you grow has been one of my life's greatest privileges, filling me with pride and offering lessons that influence my work and worldview.

To our family, thank you for your unwavering support. Your presence has been a source of strength, and your companionship invaluable in both my personal journey and this project.

To my friends and colleagues, especially those in environmental science, renewable energy, and political science, I extend my deepest gratitude. Your insights and perspectives have challenged and enriched my thinking. Engaging in debates and discussions with you has added depth and authenticity to this book.

A special thank you to those who contributed to my research on PFAS contamination in Western U.S. rivers, including the Colorado River and regional waterways, remote locations such as Hawaii and Saipan, and investigations into PFAS in food and beverages. I am especially grateful to the College of Southern Nevada students who participated in our research, their dedication was instrumental to our success. Your collaboration and efforts strengthened the scientific foundation of this work. Studying the presence of these "forever chemicals" in our rivers, drinking water, agricultural systems, and even the most isolated corners of the Pacific has been both a sobering revelation and a call to action, one that would not have been possible without the commitment of so many.

DB Sims, PhD

Forward

Per- and polyfluoroalkyl substances (PFAS), commonly known as "forever chemicals," are among the most persistent and pervasive environmental challenges of the 21st century. Once hailed for their resistance to heat, water, and oil, these synthetic compounds have become an inescapable part of modern life. From nonstick cookware to firefighting foams, waterproof clothing to food packaging, PFAS are everywhere—on the roads we drive, in the water we drink, and even in the air we breathe.

And yet, for all their convenience, PFAS have left behind an unfolding environmental catastrophe. Their nearly indestructible chemical bonds mean they do not break down naturally. They persist in soil, water, and human bodies, accumulating and spreading with every passing year. Surface waters, the lifeblood of our drinking supply and ecosystems, have become stark indicators of the broader crisis these chemicals pose to the planet.

However, the PFAS crisis is more than just an environmental disaster—it is a public health emergency. These chemicals accumulate in human tissues, disrupting key physiological functions and leading to severe long-term health consequences. They bind to proteins in the blood, distributing to vital organs like the liver, kidneys, and thyroid. Unlike other contaminants, PFAS are not metabolized and can persist in the body for years, continuously exerting their toxic effects.

The consequences are staggering. PFAS exposure has been linked to cancers, immune suppression, thyroid disorders, metabolic diseases, and reproductive harm. They alter hormone regulation, impair fetal development, and weaken immune responses. Alarmingly, they pose the greatest risks to vulnerable populations, including pregnant women, children, and the immunocompromised, increasing the likelihood of chronic illness and developmental disorders.

Making matters worse, PFAS are often absent from product labels because no federal law requires manufacturers to disclose their presence. Many companies choose not to list them due to consumer concerns and potential legal risks, even when the amounts comply with current regulations. As a result, consumers may unknowingly purchase PFAS-containing products unless they actively research specific items or manufacturer policies. This lack of transparency means the public is often unaware of the true extent of their exposure.

We have gone too far. PFAS contamination is now a global reality, with thousands of variations spreading across every continent. Once released, they travel vast distances, infiltrating drinking water, food supplies, and wildlife. These chemicals have been found in the blood of nearly every person tested and in ecosystems from Arctic ice to deep ocean currents. No one is untouched.

We have two choices: accept the consequences of living in a PFAS world or take decisive action to unwind this toxic web. The challenge is immense, but progress is possible. Countries have begun banning PFAS in consumer products. Companies are phasing them out. Scientists are developing new methods to remove them from water, soil, and air.

Yet, existing solutions remain incomplete. Filtration systems like activated carbon and reverse osmosis remove some PFAS but are costly and unsustainable. True remediation will require global cooperation, investment in safer alternatives, and a fundamental shift in chemical policy.

The next chapter of this story is still being written. Will it be one of further contamination, of rising illness, of ecosystems in decline? Or will it be one of redemption, where responsibility and resilience define our response?

This is our moment. We can either allow the mistakes of the past to dictate the health of future generations or stand up and demand a new course—one where science, justice, and sustainability lead the way.

Future generations will look back at this time. Will they see it as the moment we turned the tide, rejecting unchecked pollution and rewriting the legacy of chemical safety? Or will they see another tragedy, another warning ignored?

Chapter 1

The History of PFAS

Few chemical innovations have had as profound an impact on modern life, and the environment, as per- and polyfluoroalkyl substances (PFAS). First hailed as technological marvels, these "forever chemicals" are now known for their persistence and toxicity. Understanding the history of PFAS reveals how a groundbreaking invention transformed into a global environmental and public health crisis, with profound implications for ecosystems, human health, and corporate responsibility.

The Origin and Industrial Invention of PFAS

The story of PFAS begins in 1938 when Roy Plunkett, a young scientist at DuPont, made a serendipitous discovery that would alter the trajectory of industrial chemistry. While working on refrigerants for Freon, Plunkett stumbled upon a waxy white material with unusual properties: it was chemically inert, had a remarkably low coefficient of friction, and was highly resistant to heat (Powley et al.,

1

2005). This material, later identified as polytetrafluoroethylene (PTFE), would become the cornerstone of Teflon, a groundbreaking nonstick coating for cookware that revolutionized kitchen technology when introduced commercially in the 1940s. Teflon quickly became a household name, embodying the promise of modern chemistry to make life easier, cleaner, and more efficient.

Around the same time, 3M was making strides in the development of perfluorooctanesulfonic acid (PFOS), another PFAS compound. Initially engineered for use in military firefighting foams, PFOS was later adopted in a variety of consumer products, including stain-resistant coatings for carpets and fabrics. These products were marketed as innovative solutions to everyday problems, appealing to post-war consumers seeking convenience and durability. Both DuPont and 3M recognized that the exceptional properties of these chemicals, such as their resistance to heat, water, oil, and chemical degradation, had applications far beyond their initial uses. PFAS were hailed as "miracle chemicals," perfectly suited for the demands of a rapidly modernizing society.

The post-World War II industrial boom accelerated the proliferation of PFAS applications. By the 1950s, these chemicals were not only used in cookware and firefighting foams but had also found their way into water-repellent fabrics, food packaging, and industrial processes. Their versatility and durability became their defining characteristics, enabling innovations across industries. PFAS played a critical role in the aerospace, automotive, and electronics sectors, where their resistance to extreme conditions proved invaluable. For example, PFAS-based materials were used in the production of electrical insulation and waterproofing, supporting the rapid technological advancements of the mid-20th century.

Despite the widespread enthusiasm for PFAS, early internal research by DuPont and 3M revealed troubling findings. Scientists at both companies began documenting the unique persistence of PFAS in the environment and their tendency to accumulate in biological systems.

As early as the 1950s, DuPont researchers noted that PTFE and related compounds did not degrade under natural conditions, leading to concerns about their long-term impact on ecosystems (Kwiatkowski et al., 2020). Similarly, 3M studies in the 1960s identified bioaccumulation of PFOS in animal tissues, raising red flags about potential health risks. However, these findings were not shared with regulators or the public. Instead, the companies continued to expand PFAS production, prioritizing profit and innovation over transparency.

This silence allowed PFAS to become deeply embedded in daily life, even as their environmental and health risks remained unaddressed. Their rise from accidental discovery to global ubiquity exemplifies the dual-edged nature of scientific progress, offering transformative benefits while introducing unforeseen and enduring challenges. The decisions made during this early period laid the foundation for the widespread PFAS contamination crisis that the world faces today.

Moments in History: From Innovation to Awareness of Risks

While PFAS revolutionized manufacturing and consumer products in the mid-20th century, their environmental and health consequences became apparent much earlier than many realize. By the 1960s, internal research conducted by DuPont and 3M raised significant red flags. Company scientists discovered that PFAS were not only persistent in the environment but also bioaccumulative, meaning they could build up in the tissues of animals and humans over time (Grandjean & Clapp, 2015). Early studies indicated that PFAS did not degrade under natural conditions and could travel long distances through air and water, contaminating ecosystems far from their point of origin. Moreover, experiments with laboratory animals revealed toxic effects, including liver damage and developmental abnormalities. Despite this growing body of evidence, DuPont and 3M continued to manufacture and market PFAS-containing products, deliberately withholding these findings from the public and regulatory agencies.

The 1990s marked a pivotal shift in the narrative surrounding PFAS, as independent scientists and environmental regulators began to uncover the extent of contamination. A landmark moment occurred in 1999 when Rob Bilott, an attorney representing a West Virginia farmer, filed a class-action lawsuit against DuPont. The case, spurred by the death of the farmer's cattle, revealed extensive contamination of drinking water supplies near DuPont's Washington Works plant in Parkersburg, West Virginia. The source of the contamination was perfluorooctanoic acid (PFOA), a key PFAS compound used in the production of Teflon. Investigations linked the contaminated water to severe health issues in the local population, including cancer, birth defects, and thyroid disease (Bilott, 2019). This case, eventually settled for $343 million, exposed decades of corporate negligence and set the stage for broader public awareness of PFAS risks. The story of this legal battle was later brought to a wider audience through the 2019 film *Dark Waters*, further amplifying the call for action.

The early 2000s saw regulatory agencies finally begin to address PFAS contamination. The U.S. Environmental Protection Agency (EPA) launched investigations into PFOA and PFOS, the two most widely studied PFAS compounds, and found that both posed significant risks to human health and the environment. In response to mounting pressure, 3M voluntarily phased out PFOS in 2002, citing environmental concerns. Similarly, DuPont announced plans to cease production of PFOA by 2006 as part of an agreement with the EPA (Cousins et al., 2019). However, these actions came too late to prevent widespread contamination. By this point, PFAS had been detected in water supplies, wildlife, and human blood samples across the globe. A 2005 study conducted by the Centers for Disease Control and Prevention (CDC) found PFOA and PFOS in the blood of 98% of Americans, underscoring the pervasive nature of these chemicals.

Although the phase-out of PFOA and PFOS marked progress, it did not resolve the broader PFAS crisis. Manufacturers quickly replaced these compounds with newer "short-chain" PFAS compounds, which were initially marketed as safer alternatives but have since been found to have similar environmental persistence and potential health risks. Regulatory gaps and inconsistent enforcement have allowed PFAS contamination to continue unabated, highlighting the challenges of managing "forever chemicals" in a globalized economy.

These key moments in PFAS history illustrate the tension between innovation and responsibility. While PFAS enabled transformative advancements in technology and consumer products, the failure to address their risks in a timely and transparent manner has left a legacy of contamination that will persist for generations. This history underscores the importance of vigilance, accountability, and proactive regulation in preventing similar crises in the future.

Ubiquity of PFAS in the Products We Use and the Food Chain

PFAS have become so pervasive in modern life that they are nearly impossible to avoid. Their exceptional resistance to heat, water, grease, and chemicals has made them a go-to material for industries and manufacturers worldwide. In everyday households, PFAS are found in nonstick cookware like Teflon-coated pans, which have become kitchen staples due to their convenience and durability. Beyond the kitchen, PFAS are embedded in water-repellent clothing, stain-resistant carpets, and even cosmetics like waterproof mascaras and long-lasting lipsticks. Industrial uses are equally extensive, ranging from firefighting foams used to combat fuel fires to lubricants and chemical coatings that enhance the durability of machinery and infrastructure (Kissa, 2001).

PFAS are also ubiquitous in food packaging materials, particularly those designed to resist grease and moisture. Products like microwave popcorn bags, pizza boxes, and fast-food wrappers commonly contain PFAS, which act as barriers to prevent oil and liquids from soaking through. This use of PFAS in food packaging has introduced another pathway for human exposure, as studies have shown that these chemicals can migrate from packaging into food, especially when exposed to heat (Domingo et al., 2021). Such applications have made PFAS indispensable to modern manufacturing, but their widespread use has also led to significant environmental consequences.

The extensive use of PFAS in consumer and industrial products has resulted in contamination of air, soil, and water, creating a cycle of exposure that persists indefinitely. PFAS are released into the environment during manufacturing, use, and disposal. For instance, during industrial processes, PFAS can leach into nearby water supplies or be emitted into the air, eventually settling in soil and water bodies. Disposal of PFAS-containing products in landfills leads to leachate that can infiltrate groundwater. Firefighting foams used at military bases and airports, often in large quantities, are another significant source of PFAS contamination in surface waters and groundwater (Kwiatkowski et al., 2020).

PFAS in the Food Chain

One of the most alarming aspects of PFAS contamination is their infiltration into the food chain. These chemicals are highly mobile in the environment, making their way into rivers, lakes, and oceans, where they accumulate in aquatic organisms. Studies have consistently detected PFAS in fish and shellfish, with concentrations increasing up the food chain through a process known as biomagnification. Predatory species like tuna, salmon, and seals often exhibit higher PFAS levels, posing risks to both wildlife and humans who consume them (Domingo et al., 2021).

Agricultural systems are also not immune to PFAS contamination. When biosolids from wastewater treatment plants are used as fertilizers, they can introduce PFAS into crops. These chemicals are absorbed by plants, entering the food supply and eventually reaching consumers. Dairy products and meat have similarly been found to contain PFAS, as livestock can be exposed through contaminated water, feed, and soil. Even drinking water, a critical resource for human survival, has been widely affected. PFAS contamination has been detected in municipal water supplies across the globe, often at levels exceeding recommended safety limits (Grandjean & Clapp, 2015).

The mobility and persistence of PFAS have led to their global distribution, with traces found in even the most remote ecosystems. Research has detected PFAS in Arctic wildlife such as polar bears, seals, and seabirds, despite these regions being far from industrial centers. The presence of PFAS in such isolated areas underscores their ability to travel long distances through air and ocean currents, contaminating ecosystems on a global scale. This phenomenon highlights the inescapable reach of PFAS and the challenge of addressing their widespread contamination.

The ubiquity of PFAS in the products we use and the food we consume underscores the depth of their integration into modern life, and the complexity of mitigating their impacts. While their versatility and durability have made them indispensable to industry, these same traits have caused environmental and public health crises that are difficult to reverse. From the nonstick pan on the stove to the fish on the dinner plate, PFAS are a constant presence, silently accumulating in our bodies and ecosystems. Their journey from product to environment to food chain paints a sobering picture of how modern convenience has come at a significant cost to planetary and human health. Addressing the pervasive nature of PFAS requires a global commitment to innovation, regulation, and accountability.

Legal Cases, Corporate Accountability, and Whistleblowers

The legal and ethical dimensions of PFAS history reveal the often-overlooked consequences of unchecked corporate practices. These cases not only highlight the severe environmental and health impacts of PFAS but also illustrate the lengths to which corporations went to conceal the risks associated with their products. The 1999 class-action lawsuit against DuPont, led by attorney Robert Bilott, was a groundbreaking moment in holding corporations accountable for PFAS contamination. The case, filed on behalf of a West Virginia farmer and later expanded to represent thousands of affected residents, exposed how DuPont had knowingly discharged perfluorooctanoic acid (PFOA) into local drinking water sources near its Washington Works plant for decades. The contamination led to widespread health issues, including cancer, developmental defects, and other chronic illnesses (Bilott, 2019). Internal company documents revealed that DuPont had been aware of PFOA's toxicity and persistence since at least the 1960s but chose to suppress this information to protect its profits.

The case culminated in a $343 million settlement, which included funding for a large-scale health study to investigate the effects of PFOA exposure on the affected population. The study, known as the C8 Science Panel, confirmed a probable link between PFOA and several health conditions, including kidney cancer, testicular cancer, and pregnancy-induced hypertension. These findings formed the basis for subsequent lawsuits and regulatory action, making the Bilott case a turning point in the fight for PFAS accountability.

In 2017, DuPont and its spin-off company, Chemours, reached a $671 million settlement to resolve over 3,500 lawsuits related to PFOA contamination. This marked one of the largest environmental settlements in history, signaling the scale of the damage caused by PFAS pollution. However, critics argue that the settlement fell short of addressing the full scope of contamination and the long-term health effects on impacted communities. Similarly, 3M, the original manufacturer of perfluorooctanesulfonic acid (PFOS), faced multiple lawsuits for its role in PFOS contamination. Allegations against 3M included knowingly concealing evidence of PFOS's environmental persistence and toxicity while continuing to market products containing the chemical (Grandjean & Clapp, 2015).

Role of Whistleblowers

Whistleblowers have been instrumental in exposing the truth about PFAS contamination. Employees from DuPont and 3M leaked internal documents that provided irrefutable evidence of corporate negligence. These documents revealed that both companies had conducted studies demonstrating the harmful effects of PFOA and PFOS on animals and humans but had failed to report these findings to regulators or the public. For instance, DuPont's internal memos showed that the company had been aware of PFOA's potential to cause cancer as early as the 1980s but chose to continue its production and use (Bilott, 2019). These revelations were pivotal in building legal cases against corporations and shedding light on the broader issue of PFAS contamination.

Role of Investigative Journalism

Investigative journalists also played a critical role in uncovering the scope of PFAS pollution. Articles published in major outlets like *The New York Times*, *The Intercept*, and *The Guardian* brought the issue to national and international attention. For example, Nathaniel Rich's 2016 *New York Times Magazine* article, "The Lawyer Who Became DuPont's Worst Nightmare," provided a detailed account of Robert Bilott's two-decade legal battle against DuPont, offering readers a window into the scale of corporate negligence and the devastating impacts of PFAS pollution. Such media coverage was instrumental in educating the public and pressuring regulatory agencies to act.

Corporate Denial and Regulatory Loopholes

For decades, DuPont, 3M, and other PFAS manufacturers exploited regulatory loopholes to avoid accountability. PFAS were not classified as hazardous substances under the Resource Conservation and Recovery Act (RCRA) or the Comprehensive Environmental Response, Compensation, and Liability Act (CERCLA), also known as the Superfund law, allowing companies to discharge these chemicals with minimal oversight. Furthermore, the lack of federal regulations governing PFAS in drinking water allowed contamination to spread unchecked.

Efforts to close these regulatory gaps have been slow and contentious, largely due to the chemical industry's lobbying power. DuPont and 3M, among others, have spent millions of dollars lobbying against stricter regulations for PFAS. These tactics have delayed meaningful action, even as evidence of widespread contamination continues to mount.

Challenges of Accountability

Despite significant legal victories, challenges remain in holding corporations fully accountable for PFAS contamination. The scale of the problem is enormous, with contamination sites numbering in the thousands across the globe. Cleaning up PFAS pollution is technically and financially challenging, as these chemicals resist traditional remediation techniques. Moreover, the replacement of long-chain PFAS like PFOA and PFOS with newer "short-chain" PFAS has complicated efforts to regulate and address the issue, as these alternatives are similarly persistent and potentially harmful.

The legal battles, whistleblower revelations, and investigative efforts surrounding PFAS contamination highlight the critical need for transparency and corporate accountability in the chemical industry. While landmark cases like the Bilott lawsuit and significant settlements have set precedents for addressing PFAS pollution, the journey toward justice is far from over. The PFAS crisis underscores the importance of proactive regulation, ethical corporate practices, and vigilant public oversight to prevent similar environmental disasters in the future. As society grapples with the long-term consequences of PFAS, these cases serve as both a cautionary tale and a call to action for greater accountability and reform.

Innovation Without Accountability: Lasting Lessons of PFAS

The history of PFAS serves as a powerful cautionary tale, a narrative of innovation that prioritized convenience and profit over environmental stewardship and public health. When PFAS were first introduced, their unique properties, resistance to heat, water, and grease, revolutionized industries and transformed daily life. They were heralded as "miracle chemicals," driving advancements in cookware, firefighting, textiles, and countless other sectors. However, the very traits that made PFAS so versatile, chemical stability and durability, also made them an environmental and health hazard of unprecedented proportions.

As industries celebrated technological progress, critical questions about the long-term impacts of PFAS were left unanswered. Internal corporate studies documenting their persistence, bioaccumulation, and toxicity were deliberately withheld, allowing their unchecked use and release into the environment. This shortsightedness has led to a crisis that will endure for generations, as PFAS remain in the soil, water, and even human bloodstreams indefinitely. Today, their pervasive presence underscores the cost of prioritizing industrial innovation without fully understanding or mitigating its consequences.

The PFAS crisis is more than an environmental issue; it is a stark reminder of the dangers of unregulated industrial practices. Understanding this history is essential, not only for addressing the ongoing impacts of PFAS contamination but also for preventing similar mistakes in the future. As society continues to develop and adopt new technologies, the lessons of PFAS demand a reevaluation of how innovation is approached. Transparency, accountability, and rigorous oversight must take precedence to ensure that the pursuit of progress does not come at the expense of planetary and human health. Only by learning from the past can we navigate toward a future where technological advancement aligns with environmental and ethical responsibility.

Chapter 2

Science of PFAS

PFAS, or per- and polyfluoroalkyl substances, are often referred to as "forever chemicals" due to their extraordinary chemical stability and persistence in the environment. To understand the magnitude of their environmental and health impacts, it is crucial to examine the unique scientific properties that make PFAS both highly useful and profoundly problematic. Their durability has revolutionized industries but has also created one of the most significant contamination challenges in modern history. This chapter explores the chemistry of PFAS, the dual-edged nature of their properties, and how these chemicals move through and persist in water systems, perpetuating their global reach.

The discovery of PFAS was a marvel of modern chemistry, a breakthrough that revolutionized industries, from nonstick cookware to waterproof fabrics. Yet, beneath this scientific triumph lay an unseen consequence: a chemical legacy so persistent and pervasive that it now contaminates our water, our food, and even our bodies. What was once hailed as an innovation of convenience has become an inescapable crisis, forcing us to confront the hidden costs of progress.

Chemistry of PFAS: Why They Resist Degradation

PFAS are a class of synthetic organic compounds defined by their unique molecular structure: a chain of carbon atoms that are fully or partially bonded to fluorine atoms. This structure gives PFAS their unparalleled chemical stability. The carbon-fluorine (C-F) bond, with a bond energy of approximately 110 kcal/mol, is one of the strongest bonds known in organic chemistry, significantly stronger than the carbon-hydrogen (C-H) bond, which has a bond energy of about 83 kcal/mol (Kwiatkowski et al., 2020). This extraordinary bond strength makes PFAS highly resistant to thermal, chemical, and biological breakdown processes. As a result, they do not degrade under typical environmental conditions, leading to their characterization as "forever chemicals."

The unique chemistry of PFAS extends beyond the C-F bond. These compounds exhibit amphiphilic properties, meaning they possess both hydrophobic (water-repellent) and hydrophilic (water-attracting) characteristics. This duality is a result of their molecular structure, where the fluorinated carbon chain forms the hydrophobic tail, while the functional group at the head of the molecule interacts with water. These amphiphilic properties make PFAS highly effective in a variety of applications. For example, in nonstick coatings like Teflon, the hydrophobic tail creates a surface that resists oil and water. In firefighting foams, the hydrophilic head allows PFAS to form a stable film over burning fuel, effectively cutting off oxygen and suppressing flames. These same properties also make PFAS invaluable in textiles, where they provide water and stain resistance, and in industrial processes requiring chemical inertness.

While these characteristics make PFAS highly versatile and valuable, they also contribute to their environmental persistence. The hydrophobic tail resists dissolution in water, preventing PFAS from being easily diluted or dispersed. Meanwhile, the hydrophilic head allows them to interact with water molecules, increasing their mobility in aquatic systems. This combination enables PFAS to travel long distances through surface water and groundwater, contaminating ecosystems far from their original source (Wang et al., 2017). Additionally, their resistance to chemical, thermal, and biological breakdown means that PFAS can persist indefinitely in the environment, accumulating in soil, water, and living organisms over time.

Degradation processes that typically break down organic compounds, such as hydrolysis, photolysis, and microbial digestion, are largely ineffective against PFAS due to the strength of the C-F bond. Hydrolysis, which involves the breaking of chemical bonds through the reaction with water, cannot disrupt the C-F bond. Similarly, photolysis, the breakdown of molecules through exposure to sunlight, fails to degrade PFAS under natural environmental conditions. Microorganisms that typically digest organic matter are also unable to metabolize PFAS, leaving these compounds intact even in biologically active environments.

Traditional water treatment methods are similarly ineffective at removing PFAS from contaminated water. Techniques such as filtration, sedimentation, and biological treatment are incapable of breaking down PFAS or capturing them efficiently. Advanced water treatment methods, including reverse osmosis and activated carbon filtration, can reduce PFAS levels but are limited by high operational costs and the generation of concentrated PFAS waste that requires additional handling and disposal (Nguyen et al., 2020). These challenges underscore the urgent need for innovative and cost-effective solutions to address the growing problem of PFAS contamination in the environment.

The chemical properties of PFAS not only define their usefulness but also pose significant challenges for environmental management. Their resistance to degradation, coupled with their ability to accumulate and travel through water systems, has made them one of the most persistent and pervasive pollutants of the modern era. Understanding their chemistry is essential for developing effective strategies to mitigate their impacts and prevent further environmental and public health crises.

Properties That Make PFAS Both Useful and Dangerous

The same properties that make PFAS indispensable in industrial and consumer applications also render them profoundly dangerous to the environment and human health. The exceptional chemical and thermal stability of PFAS allows them to withstand extreme conditions that would degrade most other materials. These properties make PFAS essential in firefighting foams, where they form a stable layer over burning fuels to suppress flames, even under high-temperature conditions. Similarly, PFAS's ability to repel both water and oil has made them ideal for nonstick cookware, stain-resistant textiles, and food packaging materials, offering unmatched convenience and performance in daily life (Kissa, 2001). For decades, these qualities have driven the widespread adoption of PFAS in a variety of industries, from aerospace to consumer goods.

However, the same chemical resilience that makes PFAS valuable in products comes at an immense environmental cost. Unlike naturally occurring compounds that degrade into harmless substances over time, PFAS persist indefinitely in the environment. They are resistant to hydrolysis, photolysis, and microbial digestion, processes that typically break down organic matter. This persistence has led to the widespread contamination of air, soil, and water, where PFAS accumulate over time. As a result, these chemicals are now found in nearly every corner of the planet, from urban centers to remote ecosystems (Nguyen et al., 2020).

One of the most troubling aspects of PFAS is their bioaccumulative nature. These chemicals are absorbed by living organisms and are not easily metabolized or excreted. Instead, they accumulate in fatty tissues and organs over time, with concentrations increasing as they move up the food chain. For example, fish exposed to PFAS-contaminated water are consumed by larger predators, including humans, leading to higher levels of PFAS in top-tier species. This process, known as biomagnification, exacerbates the risks to human health, as PFAS have been linked to a range of adverse effects, including cancer, immune system suppression, and developmental issues (Domingo et al., 2021). The persistence of PFAS in human bloodstreams and tissues underscores their long-term impact on public health.

In addition to their persistence and bioaccumulation, PFAS are highly mobile in the environment. They can easily leach from products and industrial sites into surrounding soil and water, making containment efforts extraordinarily difficult. PFAS are soluble in water, enabling them to travel through rivers, lakes, and groundwater systems, often reaching locations far from their original source. Atmospheric transport also contributes to their global spread; volatile PFAS precursors, such as fluorotelomer alcohols (FTOHs), can evaporate and travel long distances through the air before transforming into more stable PFAS compounds, such as perfluorooctanoic acid (PFOA) (Grandjean & Clapp, 2015). This mobility has resulted in PFAS contamination being detected in remote areas, including the Arctic and Antarctic, where industrial activity is minimal.

The very qualities that make PFAS versatile and durable in products transform them into pollutants that resist cleanup and remediation efforts. Conventional water treatment methods, such as filtration and sedimentation, are largely ineffective at removing PFAS. Advanced methods, such as activated carbon filtration and reverse osmosis, can reduce PFAS levels but are costly and generate concentrated waste that requires further disposal. Soil remediation is similarly challenging, often requiring excavation and thermal destruction to eliminate PFAS contamination. These technical and financial barriers underscore the immense difficulty of addressing PFAS pollution on a global scale (Nguyen et al., 2020).

The dual-edged nature of PFAS highlights the broader dilemma of modern industrial innovation: how to balance technological advancement with environmental and public health safeguards. While PFAS have provided undeniable benefits in terms of product performance and industrial efficiency, their long-term consequences have revealed the risks of prioritizing convenience and profit over sustainability. Addressing the challenges posed by PFAS requires not only technical solutions but also regulatory reforms and a commitment to developing safer, more sustainable alternatives.

How PFAS Move Through the Environment and Persist in Water Systems

PFAS contamination is driven by the unique ability of these chemicals to move through and persist in environmental systems, particularly water. Once released into the environment, PFAS travel through multiple pathways, including direct discharge from industrial facilities, leaching from landfills, and the widespread use of firefighting foams. Industrial facilities producing or using PFAS in manufacturing often discharge waste containing these chemicals into nearby water bodies, contributing to contamination of rivers, lakes, and streams. Similarly, PFAS in consumer products, such as nonstick cookware and textiles, often end up in landfills, where they leach into surrounding soils and groundwater over time. Firefighting foams, extensively used at airports and military bases, are another major source of PFAS contamination, with runoff often entering surface waters or infiltrating groundwater systems. These pathways frequently lead to contamination of water resources critical for drinking water, agricultural irrigation, and ecosystem-health (Cousins et al., 2019).

Behavior of PFAS in Water Systems

In aquatic environments, PFAS exhibit unique behaviors due to their amphiphilic nature, which allows them to interact with both water and organic matter. On one hand, PFAS can dissolve and spread widely in surface water, leading to contamination of rivers, lakes, and oceans. On the other hand, PFAS compounds can bind to organic matter and sediments, where they accumulate over time. This dual behavior creates long-term sources of contamination, as PFAS in sediments can re-enter the water column through disturbance caused by dredging, flooding, or changes in environmental conditions.

Groundwater systems are particularly vulnerable to PFAS contamination due to their mobility through soil. PFAS can migrate through unsaturated soil layers and into aquifers, where they persist for decades. This contamination of groundwater is especially concerning because aquifers serve as the primary source of drinking water for many communities. Once PFAS enter these systems, they are exceedingly difficult to remove, posing significant challenges for water treatment and public health (Nguyen et al., 2020).

Atmospheric Transport of PFAS

In addition to water-based pathways, atmospheric transport plays a significant role in the global spread of PFAS, highlighting their ability to affect ecosystems far from their original sources. Volatile PFAS precursors, such as fluorotelomer alcohols (FTOHs), fluorinated sulfonamides, and other gaseous PFAS-related compounds, are released into the atmosphere during industrial processes, product use, and waste disposal. These compounds, unlike their stable counterparts, have relatively high vapor pressures, allowing them to evaporate into the atmosphere where they are carried over long distances by wind and air currents (Grandjean & Clapp, 2015).

Once airborne, these volatile PFAS precursors undergo chemical reactions in the atmosphere. Over time, they degrade into more stable perfluoroalkyl acids (PFAAs), such as perfluorooctanoic acid (PFOA) and perfluorooctanesulfonic acid (PFOS), which are known for their persistence and environmental impact. These newly formed stable compounds eventually deposit onto the Earth's surface through wet deposition (rain and snow) or dry deposition (settling of particles). This atmospheric deposition has been identified as a major contributor to PFAS contamination in areas with no local sources of pollution, including remote regions such as the Arctic and Antarctic (Young et al., 2007).

Arctic and Antarctic: A Case Study in PFAS Mobility

The detection of PFAS in remote polar regions underscores the global mobility and persistence of these chemicals. Studies have found significant levels of PFAS in Arctic snow, ice, and wildlife, including polar bears, seals, and seabirds. The presence of PFAS in these ecosystems, far removed from industrial activity, highlights the role of atmospheric transport in their distribution (Butt et al., 2010). As volatile precursors are carried by air currents to colder regions, they condense and settle due to the "cold trapping" effect, wherein lower temperatures in the Arctic and Antarctic slow the degradation process and enhance deposition (Young et al., 2007).

This contamination poses a direct threat to the health and stability of polar ecosystems. Polar bears, for example, are at the top of the Arctic food chain and have been found to have some of the highest PFAS concentrations in their tissues among wildlife species globally. This accumulation occurs through biomagnification, as PFAS are passed up the food chain from fish and seals to their predators. The impacts of PFAS on Arctic wildlife include potential immune system suppression, hormonal disruption, and reproductive challenges, raising concerns about the long-term health of these populations and the ecosystems they inhabit (Butt et al., 2010).

PFAS Atmospheric Cycling and Global Implications

The ability of PFAS to circulate through the atmosphere also means that they can undergo repeated cycles of deposition, volatilization, and re-deposition, a phenomenon known as the "grasshopper effect." This cycling further amplifies their distribution and allows PFAS to reach environments thousands of miles from their original source. For example, PFAS emitted from industrial facilities in Europe or

North America can be carried across oceans and continents, depositing in ecosystems where they were never produced or used.

Beyond the Arctic and Antarctic, PFAS have also been detected in high-altitude environments, such as mountain ranges in Europe and Asia, and in remote islands, where local human activity cannot account for their presence (Cousins et al., 2019). This global reach highlights the challenges of containing PFAS contamination and underscores the importance of international cooperation in addressing their environmental and health impacts.

Broader Impact of Atmospheric PFAS Transport

The role of atmospheric transport in the global spread of PFAS has significant implications for environmental health and regulatory efforts. As PFAS precursors degrade into stable compounds like PFOA and PFOS, their environmental persistence ensures that they continue to accumulate in water, soil, and living organisms over time. This highlights the interconnectedness of environmental systems, as PFAS released into the air ultimately contaminate water and land, compounding their impact.

Moreover, the global distribution of PFAS complicates regulatory efforts, as contamination in one region can originate from emissions in another. This underscores the need for coordinated international policies to reduce the production and release of PFAS, particularly volatile precursors that contribute to atmospheric transport. Without comprehensive global action, PFAS will continue to spread, impacting ecosystems and human populations far beyond their point of origin.

The role of atmospheric transport in the global spread of PFAS underscores their mobility, persistence, and potential for widespread contamination. The detection of PFAS in remote regions, such as the Arctic and Antarctic, serves as a stark reminder of the far-reaching consequences of industrial activity. Addressing PFAS contamination requires a holistic approach that accounts for all pathways of distribution, including atmospheric cycling. International collaboration and innovative solutions are essential to mitigating the long-term impacts of PFAS on global ecosystems and human health.

Challenges in Removing PFAS from Water Systems

The persistence of PFAS in water systems is compounded by the immense challenges of removing them once contamination has occurred. Conventional water treatment methods, such as coagulation, sedimentation, and activated sludge processes, are largely ineffective at breaking down PFAS due to their strong carbon-fluorine bonds and resistance to degradation. These traditional technologies can remove some contaminants but are insufficient for addressing PFAS, allowing these chemicals to remain in treated water and continue their environmental journey.

Advanced water treatment techniques, such as reverse osmosis, ion exchange, and activated carbon filtration, have been shown to reduce PFAS concentrations effectively. However, these methods come with significant drawbacks. They are expensive, energy-intensive, and often generate concentrated PFAS waste streams that require further disposal or destruction. Additionally, these advanced systems are not widely available or accessible, particularly in smaller or underfunded communities that rely on municipal water supplies. The technical and financial challenges of treating PFAS-contaminated water underscore the urgent need for innovative solutions, such as advanced oxidation processes or PFAS destruction technologies, to address this persistent problem (Nguyen et al., 2020).

PFAS's ability to move through and persist in environmental systems makes them a unique and enduring environmental challenge. Their mobility allows them to infiltrate surface water, groundwater, and even the atmosphere, contaminating ecosystems and water supplies on a global scale. The long-term persistence of PFAS, combined with the limitations of current water treatment technologies, highlights the critical need for innovation in addressing their contamination. Understanding the pathways and behaviors of PFAS in the environment is a necessary step toward developing effective strategies to mitigate their impacts and protect vital water resources for future generations.

Regulatory Evasion Through Structural Modifications of PFAS

Despite the growing recognition of PFAS contamination and the urgent need for mitigation, regulatory efforts have struggled to keep pace with industry tactics designed to evade oversight. One of the most concerning strategies employed by chemical manufacturers involves making minor structural modifications to PFAS molecules, such as altering chain length, changing a single fluorine atom, or modifying functional groups, to create new compounds that are technically distinct from regulated PFAS but retain similar environmental persistence and toxicity (Kwiatkowski et al., 2020). This regulatory loophole, often referred to as "regrettable substitution," enables the continued production and use of PFAS variants that evade restrictions while maintaining their hazardous environmental footprint (Cousins et al., 2019).

The Transition to Short-Chain PFAS

A key example of this regulatory evasion is the industry's shift from long-chain PFAS, such as perfluorooctanoic acid (PFOA) and perfluorooctanesulfonic acid (PFOS), to short-chain alternatives like perfluorobutanoic acid (PFBA) and perfluorobutanesulfonic acid (PFBS). This transition was prompted by regulatory actions in the United States and Europe following mounting evidence linking long-chain PFAS to cancer, immune system suppression, and other adverse health effects (Grandjean & Clapp, 2015). However, while short-chain PFAS are less bioaccumulative, they remain highly persistent in the environment and exhibit similar toxicity concerns, raising questions about whether these replacements are truly safer or merely a means to circumvent regulation (Blum et al., 2021).

Fluorine Substitutions and Functional Group Modifications

Beyond chain length adjustments, the industry has also introduced PFAS derivatives that feature subtle modifications to their chemical structures, effectively skirting regulatory classifications. By replacing a single fluorine atom or altering the functional head group, companies can produce compounds that avoid restrictions while retaining the desirable properties of PFAS, such as water and oil repellency, heat resistance, and chemical inertness (Cousins et al., 2019). Many of these "alternative" PFAS compounds eventually degrade into perfluoroalkyl acids (PFAAs), which persist indefinitely in the environment, reinforcing the same contamination cycle that regulators aimed to prevent (Wang et al., 2017).

The Struggle for Comprehensive PFAS Regulation

The ability of industry to introduce slightly modified PFAS faster than regulatory agencies can evaluate them has resulted in a frustrating game of regulatory catch-up. The U.S. Environmental Protection Agency (EPA) and the European Chemicals Agency (ECHA) have faced significant challenges in tracking the ever-expanding number of PFAS variants, many of which have been introduced without comprehensive toxicological assessments (Grandjean & Clapp, 2015). Traditional chemical regulations operate on a compound-by-compound basis, meaning that each new PFAS must undergo separate risk assessment and regulatory approval, allowing industry to continuously exploit this loophole.

Recognizing this limitation, regulatory bodies have begun to advocate for a class-based approach to PFAS regulation. The European Union, for example, has proposed broad restrictions under the Registration, Evaluation, Authorisation and Restriction of Chemicals (REACH) framework that would encompass all PFAS rather than addressing them individually (Cousins et al., 2019). However, industry opposition and legal challenges have slowed the implementation of such comprehensive measures, highlighting the difficulty of enforcing meaningful restrictions on a group of chemicals that remain essential to many industrial applications.

The Need for Proactive Policy Solutions

The science of PFAS underscores a profound paradox: the very properties that have revolutionized industries and enhanced modern convenience have also unleashed an environmental crisis of unprecedented scale. Their chemical resilience, bioaccumulative nature, and ability to persist and travel through air and water systems have transformed PFAS from technological marvels into some of the most insidious pollutants in history. While their durability has driven industrial and consumer advancements, it has also made them nearly impossible to remove from the environment, leading to widespread contamination of water, soil, and living organisms.

Compounding this challenge is the tactical evasion employed by the PFAS industry, which continually introduces slightly modified variants to sidestep regulations while maintaining their hazardous environmental footprint. The introduction of short-chain PFAS and subtle molecular alterations has allowed manufacturers to avoid bans and restrictions, perpetuating contamination despite growing scientific consensus on their risks. This cycle of pollution, regulatory loopholes, and industry adaptation highlights the broader challenges of chemical oversight in an era of rapid innovation.

To break this cycle, regulators must move beyond reactive, compound-specific bans and adopt a preventative, class-based approach that prohibits all PFAS variants unless proven safe. Without such a shift, environmental protections will remain one step behind industry tactics, and the long-term risks to human health and ecosystems will continue to escalate.

As society grapples with the consequences of these "forever chemicals," understanding their science provides a foundation for action—one that must combine innovation, regulation, **and collaboration** to forge a sustainable path forward. The burden now rests on the scientific community, policymakers, and industry leaders to pioneer solutions that prevent future contamination while ensuring that technological progress and environmental responsibility go hand in hand. Only through decisive and comprehensive action can we hope to mitigate the global legacy of PFAS and protect vital resources for future generations.

Chapter 3

Major Sources of PFAS Contamination

PFAS, or per- and polyfluoroalkyl substances, have infiltrated nearly every aspect of modern life due to their unique chemical properties, including water resistance, grease resistance, and thermal stability. However, these same characteristics make them persistent pollutants that accumulate in the environment and human bodies. PFAS contamination primarily stems from industrial processes, consumer goods, and waste management systems. Understanding the major sources of PFAS exposure is critical for mitigating risks and preventing further contamination.

Despite their widespread use, PFAS are often missing from product labels. There is no federal law requiring manufacturers to disclose their presence, even though these chemicals may be included in everything from food packaging to personal care products. Many companies choose not to list PFAS due to consumer concerns, potential legal ramifications, and the fact that regulatory thresholds for "safe" levels remain a subject of ongoing debate. This lack of transparency leaves consumers unaware of their exposure unless they actively research specific products or manufacturer policies. As a result, PFAS continue to permeate everyday life with little public awareness or oversight.

Industrial Sources of PFAS

Many industries have historically relied on PFAS due to their exceptional chemical properties, including water repellency, heat resistance, chemical stability, and nonstick capabilities. These characteristics have made PFAS integral to a wide range of applications, from manufacturing to consumer products. However, the widespread use of PFAS has led to significant environmental contamination, with industrial discharges being a primary contributor. Below are some of the largest industrial sources of PFAS contamination and their associated environmental impacts. Some of the largest industrial sources of PFAS contamination include:

Chemical Manufacturing and Processing Plants

Chemical manufacturing facilities have historically been among the largest sources of per- and polyfluoroalkyl substances (PFAS) contamination, with companies like 3M and DuPont leading the production of these persistent chemicals. Two of the most widely used PFAS compounds, perfluorooctanoic acid (PFOA) and perfluorooctanesulfonic acid (PFOS), were extensively incorporated into a range of industrial and consumer applications, including nonstick coatings, water-resistant textiles, firefighting foams, and stain repellents (Bilott, 2019).

One of the most well-documented cases of PFAS pollution stemming from chemical manufacturing is the Ohio River Valley contamination, where industrial wastewater discharges from DuPont's Washington Works plant led to widespread PFAS accumulation in local waterways and drinking water supplies. This contamination ultimately resulted in legal action and regulatory scrutiny, shedding light on the long-term environmental and human health risks associated with PFAS exposure. Similar cases have been observed near other PFAS manufacturing plants, where improper disposal practices, leaks, and accidental spills have led to groundwater and surface water contamination.

While regulatory efforts and public pressure have led to the phase-out of long-chain PFAS compounds such as PFOA and PFOS in many regions, manufacturers have shifted toward alternative short-chain PFAS chemicals. One such example is the GenX family of chemicals, which were introduced as replacements for legacy PFAS but have been found to persist in the environment and exhibit similar toxicological concerns (Grandjean & Clapp, 2015). Despite industry claims of reduced bioaccumulation potential, emerging research suggests that these alternative compounds still pose significant risks to human health and ecosystems, fueling ongoing debates over the adequacy of current regulatory frameworks and the need for stricter oversight of PFAS production and emissions.

As chemical manufacturers continue to develop new PFAS alternatives, concerns remain about their environmental fate, potential toxicity, and regulatory oversight. The persistence and mobility of these compounds underscore the necessity for improved industrial wastewater treatment, stronger environmental monitoring programs, and comprehensive policies to prevent further PFAS contamination in communities worldwide.

Firefighting Foams (AFFF – Aqueous Film-Forming Foam)

Aqueous Film-Forming Foam (AFFF) has been a staple in firefighting operations for decades, particularly at military installations, airports, chemical plants, and industrial facilities where flammable liquid fires pose significant hazards. Developed in the 1960s, AFFF was designed for rapid fire suppression, forming a thin, oxygen-blocking film over burning fuel surfaces. Its effectiveness made it indispensable in high-risk environments, but its widespread use also led to significant environmental contamination due to the presence of per- and polyfluoroalkyl substances (PFAS), which are highly persistent in the environment (Schultz et al., 2006).

One of the most concerning aspects of AFFF use is its role as a primary source of PFAS contamination in groundwater and drinking water supplies. During routine firefighter training exercises, accidental spills, and emergency fire suppression events, large quantities of AFFF have been released directly onto the ground, allowing PFAS to seep into the soil, migrate through subsurface layers, and accumulate in underlying aquifers. Unlike many other pollutants, PFAS do not readily break down in the environment, leading to long-term contamination that can persist for decades. This issue is particularly pronounced at sites with frequent AFFF use, such as military airfields and civilian airports, where repeated applications have led to the widespread presence of PFAS in local water supplies.

The contamination of groundwater near U.S. military installations has prompted extensive investigations and multi-billion-dollar remediation efforts. For example, at Peterson Air Force Base in Colorado, extensive PFAS contamination from AFFF use affected drinking water sources for surrounding communities, leading to significant health concerns and legal actions. Similarly, at Naval Air Station Pensacola in Florida, PFAS contamination from decades of AFFF use has necessitated costly environmental cleanups and increased regulatory scrutiny (Domingo et al., 2021). Other affected sites include military bases in Michigan, New York, and California, where water testing has revealed PFAS levels exceeding federal health advisory limits.

In response to these growing concerns, regulatory agencies and military organizations have been working to phase out the use of PFAS-containing AFFF and transition to fluorine-free firefighting foams. The Department of Defense (DoD) has committed to finding safer alternatives, and the Federal Aviation Administration (FAA) has also mandated the eventual replacement of AFFF at commercial airports. However, the challenge remains that PFAS-free alternatives must still meet stringent fire suppression standards, and legacy contamination from historical AFFF use continues to pose significant remediation challenges.

The persistence of PFAS from firefighting foams highlights the urgent need for improved containment strategies, enhanced monitoring of contaminated sites, and stronger regulatory measures to prevent future environmental and public health risks. The scale of the problem underscores the importance of ongoing research into effective PFAS removal technologies, including advanced filtration, soil remediation, and innovative destruction methods, to address the widespread impact of AFFF contamination.

Textile and Paper Manufacturing

The textile and paper manufacturing industries have been significant contributors to per- and polyfluoroalkyl substances (PFAS) contamination due to their widespread use of these chemicals in producing waterproof, stain-resistant, and grease-repellent products. PFAS have been incorporated into industrial textile manufacturing for decades, enhancing the durability and performance of everyday consumer goods such as outdoor apparel, carpets, upholstery, and home furnishings. These chemicals provide essential properties such as water repellency in rain jackets, stain resistance in furniture and automotive interiors, and oil resistance in uniforms used in various industries. However, the widespread application of PFAS in textiles has led to concerns over environmental contamination, particularly through industrial wastewater discharges and consumer product degradation over time.

In addition to textiles, PFAS are heavily used in the paper manufacturing industry, particularly in the production of grease-resistant food packaging. Factories that produce fast-food wrappers, paper bowls, pizza boxes, microwave popcorn bags, and bakery papers often apply PFAS-based coatings to prevent oils and moisture from seeping through the packaging (Nguyen et al., 2020). These coatings allow fast-food packaging to withstand hot, greasy foods while maintaining structural integrity. However, studies have shown that PFAS can leach from packaging into food, leading to direct human exposure, as well as contribute to environmental pollution when discarded materials enter landfills or incinerators. The breakdown of these materials in waste streams has been identified as a significant source of PFAS in landfill leachate, which can eventually contaminate groundwater supplies.

One of the most well-documented cases of PFAS contamination from the textile and paper industries is the Cape Fear River Basin in North Carolina, where industrial discharges from paper mills and textile factories have led to the widespread presence of PFAS in local waterways. The contamination has impacted drinking water supplies for thousands of residents, with studies revealing high concentrations of GenX and other PFAS compounds linked to industrial manufacturing activities. The case has prompted increased scrutiny of PFAS emissions from manufacturing facilities, leading to legal actions, regulatory investigations, and demands for stricter pollution control measures.

The release of PFAS-laden wastewater from textile and paper manufacturing plants has become an urgent environmental concern, as traditional wastewater treatment facilities are not equipped to effectively remove these persistent chemicals. As a result, PFAS continue to accumulate in rivers, lakes, and drinking water supplies, posing long-term risks to aquatic ecosystems and human health. Efforts to phase out PFAS use in textiles and paper products have gained traction in recent years, with some manufacturers exploring alternative, non-fluorinated coatings for waterproofing and grease resistance. However, regulatory actions remain fragmented, and the legacy of PFAS pollution from decades of manufacturing continues to present significant remediation challenges.

The ongoing presence of PFAS in textile and paper manufacturing underscores the need for more stringent industrial regulations, improved wastewater treatment technologies, and increased transparency from manufacturers regarding the use of PFAS in consumer products. As research continues to uncover the full extent of PFAS-related health and environmental risks, policymakers and industry leaders must work collaboratively to develop safer alternatives and minimize further contamination.

Consumer Products Containing PFAS

Beyond industrial sources, per- and polyfluoroalkyl substances (PFAS) are found in a wide range of everyday consumer goods, resulting in continuous human exposure through inhalation, ingestion, and dermal absorption. PFAS are valued for their waterproof, grease-resistant, stain-resistant, and nonstick properties, making them prevalent in household products, clothing, cookware, and packaging. However, due to their environmental persistence and potential health risks, the presence of PFAS in consumer goods has become an increasing concern.

Nonstick Cookware (Teflon Products)

One of the most well-known PFAS applications is nonstick cookware, where polytetrafluoroethylene (PTFE), the chemical behind Teflon, is used to coat frying pans, bakeware, and kitchen utensils to prevent food from sticking. This innovation revolutionized home cooking, offering convenience and easy cleaning. However, research has raised concerns about PFAS-related health risks, particularly when cookware is overheated or damaged, potentially releasing toxic fumes into the air.

In response to public scrutiny, manufacturers like DuPont (now Chemours) phased out PFOA-based Teflon production in the early 2000s. However, this did not eliminate PFAS from nonstick coatings; instead, newer alternative short-chain PFAS have been introduced. While these chemicals are marketed as safer, studies suggest they may pose similar health and environmental risks, as they remain persistent and bioaccumulative (Bilott, 2019). The continued use of PFAS in cookware highlights the challenge of finding truly safe alternatives while maintaining nonstick performance.

Water-Resistant Clothing and Outdoor Gear

PFAS are widely used in outdoor apparel and gear, particularly in rain jackets, ski pants, hiking boots, and camping equipment, to provide waterproofing and stain resistance. Gore-Tex, a well-known outdoor brand, historically relied on PFAS-based membranes to make its products highly water-resistant while remaining breathable. The performance benefits of PFAS-treated gear have made them a staple for outdoor enthusiasts, but concerns over environmental contamination have led some companies to explore PFAS-free alternatives.

The primary concern with PFAS in textiles is that these chemicals leach into the environment over time through:

- **Laundering:** Washing PFAS-treated clothing releases the chemicals into wastewater, which can ultimately contaminate rivers, lakes, and drinking water sources.

- **Wear and Tear:** As clothing breaks down, PFAS particles can enter the air and dust, leading to indoor and outdoor contamination.

In response to growing concerns, brands such as Patagonia, REI, and The North Face have pledged to phase out PFAS in outdoor clothing, but fully PFAS-free alternatives remain a work in progress due to durability and performance challenges (Grandjean & Clapp, 2015).

Stain-Resistant Carpets and Furniture

Many carpets, rugs, sofas, mattresses, and upholstered furniture are treated with PFAS-based stain-resistant coatings to repel spills and prevent permanent stains. Scotchgard and Stainmaster are two well-known brands that have historically used PFAS formulations, which have come under scrutiny due to their long-lasting environmental persistence and ability to accumulate in human blood.

Indoor PFAS exposure from stain-resistant treatments occurs primarily through:

- **Household Dust:** Over time, PFAS coatings degrade, releasing microscopic particles that settle into indoor dust, which can then be inhaled or ingested.

- **Direct Contact:** Sitting or lying on PFAS-treated furniture may result in dermal absorption of PFAS.

- **Outdoor & Gear Waterproofing Sprays:** These are used on furniture for water and stain proofing and shoes, fabrics, and many other household uses.

- **Airborne Release:** PFAS-treated textiles can slowly emit volatile fluorinated compounds into indoor air, contributing to prolonged exposure.

- **Cleaning products**: Used in stain removers, glass cleaners, and grease-cutting sprays

- **Car wax & cleaners**: PFAS creates water-repellent and shine-enhancing effects

- **Paints & sealants:** Industrial and household paints often contain PFAS for durability

As awareness of PFAS-related health concerns grows, some furniture manufacturers have shifted toward PFAS-free stain-resistant treatments. However, because of PFAS's exceptional durability, older carpets and furniture continue to serve as long-term sources of indoor contamination even after regulatory changes.

Food Packaging

One of the most direct routes of PFAS exposure comes from food packaging, where these chemicals are used to create grease-resistant barriers that prevent oils and liquids from soaking through packaging materials. Common sources of PFAS-containing food packaging include:

- **Fast-food wrappers and bowels** (e.g., burger/burrito wrappers, paper bowels, containers, fry containers)

- **Microwave popcorn bags**, where PFAS prevent oils from leaking out of the bag

- **Pizza boxes**, to keep grease from soaking through the cardboard

- **Paper plates and takeout containers**, which require oil- and water-resistant coatings

- **Candy wrappers**: Common in chocolate bar wrappers and chip bags

- **Processed meats**: Some deli papers and meat-packaging films contain PFAS

- **Dairy products**: PFAS have been detected in milk and dairy from contaminated water sources

- **Seafood:** Fish and shellfish bioaccumulate PFAS, especially those caught in polluted waters

- **Tea bags and other drink delivery devises**: Some brands coat their bags with PFAS for heat resistance

- **Bottled water**: Some brands have tested positive for PFAS contamination

Several studies have found that PFAS can migrate from packaging into food, particularly when exposed to heat, grease, or acidic ingredients. This means that eating food wrapped in PFAS-treated packaging can result in direct ingestion of these chemicals, contributing to long-term accumulation in the human body (Domingo et al., 2021).

In recent years, some governments and companies have taken steps to reduce PFAS use in food packaging, with major fast-food chains like McDonald's, Starbucks, and Chipotle pledging to phase out PFAS in their packaging materials. However, the transition to safer, non-fluorinated alternatives remains a challenge, and legacy PFAS contamination from food packaging waste continues to impact the environment.

The presence of PFAS in everyday consumer products highlights the widespread and persistent nature of these chemicals. While industry efforts to replace long-chain PFAS with newer alternatives are ongoing, concerns remain about the safety and environmental impact of these replacements. As regulatory pressures increase and consumer awareness grows, there is a push for greater transparency from manufacturers, stricter regulations on PFAS use, and the development of truly safe alternatives to reduce long-term exposure risks.

Personal Care Products and Cosmetics

The use of per- and polyfluoroalkyl substances (PFAS) in personal care products and cosmetics is a growing concern due to the potential for direct human exposure through skin absorption, inhalation, and ingestion. PFAS are commonly added to beauty and hygiene products to enhance water resistance, durability, and spreadability, making them highly desirable for long-lasting formulations. However, due to their persistent nature, these chemicals can accumulate in the body over time and pose potential health risks. Here are some examples:

- **Shampoo & conditioner:** Some hair care products use PFAS for smoothness and frizz control

- **Toothpaste:** Certain fluoride-containing formulas use PFAS-based compounds

- **Deodorants:** Some antiperspirants contain PFAS for sweat resistance

- **Waterproof mascara and eyeliner:** PFAS to resist smudging and moisture.

- **Waterproof makeup:** Lipsticks and foundations often contain PFAS for long-wear properties

- **Long-wear foundation and concealers:** PFAS to create a smooth, sweat-resistant finish.

- **Lipsticks and lip balms:** PFAS enhance texture and extend wear time.

- **Pressed and loose powders:** PFAS improve blending and adherence to the skin.

- **Blush & Highlighter:** PFAS enhances bendability and water resistance

- **Toilet tissue and paper towel:** Manufacturers add PFAS to wood pulp during manufacturing

- **Creams and lotions:** Found in some formulas for smooth application and hydration lock

- **Chapstick and lip balm:** PFAS are used to make them more spreadable, waterproof, and long-lasting

- **Disposable diapers:** Waterproofing layers, absorbent core materials, and manufacturing contamination

A 2021 study by the University of Notre Dame by Peaslee et al. (2021) found that over half of the tested cosmetics contained PFAS, with many products failing to disclose their presence on ingredient labels. This lack of transparency makes it difficult for consumers to avoid exposure, leading to calls for stricter regulations on PFAS use in the beauty industry.

PFAS in Hygiene and Skincare Products

PFAS contamination extends beyond cosmetics into everyday personal care items, including:

- **Sunscreens**: Contain PFAS to enhance waterproof and sweatproof properties.

- **Dental floss**: Those marketed as glide or non-stick, which often use PFAS-based coatings to slide easily between teeth.

- **Moisturizers and lotions:** PFAS creates a silky texture and improves absorption.

- **Hair Spray & Styling Products**: Helps maintain hold and humidity resistance

The primary risks of PFAS exposure from personal care and hygiene products include dermal absorption, inhalation (from sprays and powders), and accidental ingestion (from lip products and dental floss). While some companies have committed to phasing out PFAS in cosmetics, these chemicals persist in many mainstream beauty and hygiene products, increasing the likelihood of long-term exposure (Schultz et al., 2006).

PFAS in Feminine Hygiene Products

Feminine hygiene products are meant to support health and well-being, yet many contain PFAS (per- and polyfluoroalkyl substances), the same toxic "forever chemicals" found in industrial waste, nonstick cookware, and waterproof clothing. These chemicals, known for their stain-resistant, moisture-repellent, and durability-enhancing properties, have made their way into products that millions of women use daily, often without any disclosure from manufacturers.

PFAS are found in a variety of Feminine hygiene products, including:

- **Menstrual Pads:** Some brands use PFAS coatings to create moisture resistance and prevent leaks.

- **Tampons:** PFAS have been found in certain tampons, raising concerns about internal exposure.

- **Panty Liners:** Added for **moisture control and absorbency**, but at the cost of potential chemical exposure.

- **Period Underwear:** Some "leak-proof" and moisture-wicking underwear brands contain PFAS to repel liquid.

- **Incontinence Pads & Products:** PFAS coatings help maintain dryness but may pose health risks over long-term use.

- **Feminine Wipes:** Used for freshness but may contain PFAS as a preservative or stabilizer.

Cleaning Products

PFAS are also widely used in household and industrial cleaning products, where they provide water and stain resistance, durability, and grease-repelling properties. These chemicals help enhance product effectiveness by preventing dirt and oil from reattaching to surfaces after cleaning. However, because PFAS do not break down easily, their presence in cleaning supplies contributes to indoor contamination and long-term environmental persistence.

PFAS are found in a variety of cleaning and maintenance products, including:

- **Stain-resistant sprays** (e.g., fabric protectors): PFAS create a waterproof and stain-resistant barrier on fabrics, carpets, and furniture.

- **Carpet and upholstery cleaners:** PFAS help repel spills and prevent stains from setting.

- **Glass and surface cleaners**: PFAS provide streak resistance and a nonstick finish, reducing buildup.

- **Multipurpose sprays**: PFAS improve grease-cutting ability and allow cleaning products to spread evenly.

- **Oven and grill cleaners:** PFAS enhance heat resistance and help break down baked-on grease.

- **Toilet bowl and bathroom cleaners:** PFAS prevents water stains and soap scum buildup.

- **Floor waxes and sealants**: PFAS provide water and stain resistance, increasing longevity.

- **Wood and furniture polishes**: PFAS create a protective barrier against spills and moisture.

- **Mold and mildew-resistant sprays:** PFAS prevent moisture absorption, reducing mold growth.

PFAS in cleaning products pose serious health and environmental risks due to their persistence, bioaccumulation, and toxic effects. These chemicals can contaminate indoor air, surfaces, and wastewater, leading to long-term exposure linked to hormonal disruption, immune suppression, and increased cancer risks. Additionally, their resistance to breakdown means they contribute to widespread water pollution, affecting both ecosystems and drinking supplies.

Moreover, the risks associated with PFAS in cleaning products also stem from:

- **Surface Residues**: PFAS can remain on cleaned surfaces for extended periods, leading to potential dermal exposure through regular contact.

- **Airborne Exposure**: Spraying PFAS-containing cleaning products aerosolizes these chemicals, allowing them to be inhaled.

- **Wastewater Contamination**: When cleaning products are washed down the drain, PFAS enter sewage systems and municipal wastewater treatment plants, which often lack the capacity to effectively remove them before water is discharged into rivers and lakes.

Efforts to eliminate PFAS from cleaning products have gained momentum in recent years, with some brands now offering PFAS-free alternatives. However, many households and industrial cleaners still contain these chemicals, and regulatory oversight remains limited.

The presence of PFAS in personal care products, cosmetics, and cleaning supplies represents a significant and direct exposure route for consumers. As these chemicals continue to be absorbed through the skin, inhaled, or ingested, concerns over long-term health effects and environmental persistence are growing. While some companies have pledged to reduce PFAS use, regulatory gaps and lack of labeling transparency make it difficult for consumers to avoid them entirely. Increased awareness, stricter regulations, and safer product formulations are essential to mitigating the risks associated with these everyday exposures.

Waste Management and PFAS Disposal

The improper disposal of PFAS-containing products significantly contributes to environmental contamination, as these chemicals do not readily break down in nature. PFAS waste enters the environment through multiple pathways, including landfills, incineration, and wastewater treatment systems, creating long-term pollution that affects soil, water, air, and food supplies. Managing PFAS disposal remains a critical challenge due to their resistance to conventional degradation processes and the ongoing production of PFAS-containing materials.

Landfills and Leachate

Landfills serve as a major repository for PFAS-containing waste, including discarded textiles, food packaging, nonstick cookware, personal care products, firefighting foams, and electronics. Because PFAS are engineered for durability and resistance to breakdown, they persist for decades or longer in landfill environments, gradually leaching into soil and groundwater.

Key concerns associated with PFAS in landfills:

- **Slow Degradation**: Unlike organic waste, PFAS-containing items do not decompose fully, remaining in the landfill indefinitely.

- **Leachate Contamination**: As rainwater filters through waste, it dissolves and mobilizes PFAS, creating PFAS-laden leachate that can seep into groundwater or be discharged into surface water sources.

- **Leachate Treatment Limitations**: Many landfill leachate treatment systems are not designed to effectively remove PFAS, allowing these chemicals to enter nearby water bodies untreated (Nguyen et al., 2020).

- **Airborne Emissions:** PFAS compounds can volatilize from landfill waste or be released through landfill gas emissions, contributing to atmospheric contamination and potential long-range transport.

- **Bioaccumulation in Wildlife:** PFAS that escape from landfills into surrounding ecosystems can accumulate in plants, animals, and aquatic organisms, leading to biomagnification in the food chain and potential human exposure.

Case Study: PFAS in Landfill Leachate

Studies have found high concentrations of PFAS in landfill leachate in multiple locations across the U.S. and Europe, with contaminated leachate being discharged into local rivers and lakes. Some municipal wastewater treatment plants attempt to treat landfill leachate before releasing it, but because PFAS resist conventional treatment processes, they persist in wastewater effluent and spread downstream, impacting drinking water supplies.

Efforts to mitigate PFAS leachate contamination include:

- Enhanced landfill liners and leachate collection systems to prevent groundwater infiltration.

- Advanced treatment technologies, such as Granular Activated Carbon (GAC), ion exchange resins, and high-pressure membrane filtration (reverse osmosis), though these are costly and require frequent maintenance.

- Increased regulation on PFAS disposal to reduce the volume of PFAS-laden waste entering landfills.

Incineration Challenges

Incineration has been proposed as a method to destroy PFAS, but studies suggest significant challenges in fully eliminating these chemicals. Many high-temperature incinerators operate at temperatures between 800°C and 1000°C (1472°F–1832°F), which are often insufficient to completely break down PFAS compounds. Instead, incomplete combustion can lead to:

- The release of toxic PFAS byproducts into the atmosphere, which can travel long distances before depositing onto land and water surfaces.

- Reformation of PFAS chemicals, where thermal degradation creates smaller, yet still persistent, fluorinated compounds.

- Contamination of incinerator ash, which, if landfilled, can continue the PFAS cycle.

Regulatory and Research Developments

- Some hazardous waste incinerators are now required to operate at temperatures above 1100°C (2012°F) to improve PFAS destruction efficiency (Domingo et al., 2021).

- Research is underway to develop plasma-based destruction technologies that can break PFAS bonds at the atomic level, offering a potential solution for complete PFAS degradation (Grandjean & Clapp, 2015).

- The U.S. Environmental Protection Agency (EPA) is currently assessing the effectiveness of high-temperature incineration for PFAS waste disposal, as concerns remain about airborne emissions and secondary contamination.

Due to uncertainties in PFAS incineration effectiveness, some states and countries have placed moratoriums on burning PFAS waste until more reliable destruction methods are confirmed.

Wastewater Treatment Plants

Municipal wastewater treatment plants (WWTPs) are not equipped to remove or break down PFAS, leading to continuous contamination of water supplies. PFAS enter wastewater systems through:

- Industrial discharges from manufacturing plants.

- Household wastewater, including residues from shampoos, cosmetics, nonstick cookware, and laundry wastewater containing PFAS-treated textiles.

- Runoff from firefighting foam at military bases, airports, and training facilities.

Because PFAS are highly water-soluble, they pass through conventional treatment processes largely unchanged, leading to:

- **PFAS-contaminated effluent**: Treated wastewater discharged into rivers and lakes still contains PFAS, impacting drinking water sources and aquatic life (Sims et al., 2025).

- **Sewage sludge (biosolids) contamination**: PFAS accumulate in sludge, which is often applied as agricultural fertilizer, leading to contamination of soil, crops, and livestock (Domingo et al., 2021).

Case Study: Biosolids and Agricultural Contamination

In Maine, dairy farms using PFAS-contaminated biosolids as fertilizer experienced high PFAS levels in soil and milk, leading to livestock contamination and farm closures. Similar cases have been reported in Michigan, Wisconsin, and Germany, prompting bans on biosolid application in some regions.

Emerging PFAS Treatment Technologies

To address PFAS persistence in wastewater, researchers and engineers are exploring advanced treatment technologies, including:

- **Granular Activated Carbon** (GAC): Effective for removing some PFAS but requires frequent filter replacement and disposal of PFAS-laden carbon.

- **Ion Exchange Resins**: Can selectively capture certain PFAS compounds, but efficiency varies based on water chemistry.

- **Reverse Osmosis (RO)**: Highly effective for removing PFAS but produces concentrated PFAS waste, which requires further disposal solutions.

- **Electrochemical and Plasma Treatment**: Emerging destruction methods that may completely degrade PFAS molecules but require further testing for large-scale implementation.

Urgent Challenge of PFAS Contamination and Waste Management

PFAS contamination originates from a vast network of industrial processes, consumer products, and waste disposal systems, making exposure nearly unavoidable in daily life. These persistent chemicals are found in household goods, food packaging, textiles, and firefighting foams, leading to widespread environmental accumulation. Legacy pollution from industrial discharges, military firefighting foams, and landfills continues to infiltrate water sources worldwide, posing long-term risks to both ecosystems and human health. Given the persistence and mobility of PFAS, immediate action is essential to regulate their use, develop safer alternatives, and implement more effective waste management strategies.

The disposal and management of PFAS waste remain one of the most pressing environmental challenges of our time. Landfills, incineration, and wastewater treatment plants not only fail to contain PFAS but often contribute to their continued spread, underscoring the urgent need for better disposal solutions. Without decisive action, these chemicals will continue to accumulate in drinking water, food supplies, and the broader environment, exacerbating their already significant health and ecological impacts.

To mitigate PFAS contamination and prevent further harm, a multi-faceted approach is necessary, one that integrates regulatory oversight, industry accountability, and consumer awareness. Key actions for improved PFAS waste management include:

- Developing safer destruction technologies, such as plasma-based treatments, to eliminate PFAS at the molecular level rather than merely redistributing them.

- Enhancing landfill containment strategies to prevent PFAS-laden leachate from infiltrating groundwater and contaminating drinking water supplies.

- Restricting biosolid use in agriculture, as PFAS-contaminated sewage sludge has been shown to introduce these chemicals into the food chain, impacting soil, crops, and livestock.

- Investing in advanced wastewater treatment solutions, such as granular activated carbon (GAC), ion exchange resins, and high-pressure membrane filtration, to remove even low concentrations of PFAS before discharge into surface waters.

- Strengthening regulations and monitoring to prevent continued PFAS pollution from both industrial waste streams and consumer products.

The Hidden Threat: Why PFAS Are Missing from Labels

One of the most alarming aspects of PFAS contamination is that consumers often have no way of knowing when they are exposed. Despite the widespread use of these chemicals in everyday products, there is currently no federal law requiring manufacturers to disclose their presence. This regulatory gap allows companies to omit PFAS from ingredient lists, leaving consumers in the dark about their potential exposure.

Manufacturers have little incentive to voluntarily disclose PFAS use, especially given growing concerns about their environmental and health impacts. Even when PFAS are used within "safe" regulatory limits, companies may fear consumer backlash or legal challenges if their products are explicitly labeled as containing these "forever chemicals." As a result, PFAS can be found in everything from nonstick cookware to water-resistant clothing, food packaging, and personal care products—without any clear indication on the label.

For consumers seeking to avoid PFAS, navigating product labels requires vigilance. While most products do not explicitly list PFAS, certain keywords can serve as warning signs. Items marked **as** waterproof, stain-resistant, or grease-repellent often contain PFAS-based coatings. Similarly, ingredient lists that mention fluorinated compounds may indicate the presence of these chemicals. Consumers can also turn to third-party certifications**, such as** those from environmental organizations, that specifically test for and prohibit PFAS.

As awareness of PFAS contamination grows, pressure is mounting for stricter regulations and transparency. Some companies are beginning to phase out PFAS and voluntarily disclose their efforts, but until comprehensive labeling requirements are enacted, consumers must take an active role in researching the products they bring into their homes.

The scale of the PFAS crisis demands collaborative efforts between scientists, regulators, industries, and consumers. While research into safer alternatives and improved remediation technologies continues, decisive policy action is necessary to curb future contamination. Without swift intervention and long-term regulatory strategies, PFAS pollution will remain an escalating environmental and public health catastrophe. Addressing this challenge is not just about cleaning up the past but ensuring a safer and healthier future for generations to come.

However, the lack of labeling remains a direct and immediate threat to consumers. Without transparency, individuals are unknowingly exposed to these harmful chemicals in everyday products, stripping them of the ability to make informed choices about their health and safety. Until comprehensive regulations require manufacturers to disclose PFAS content, consumers remain at risk, navigating an invisible danger with little guidance. True progress will require not just policy reforms and corporate accountability but also a commitment to empowering the public with the knowledge and tools to protect themselves and future generations.

Chapter 4

PFAS in the Environment

As PFAS has become an enduring feature of the global environment, infiltrating ecosystems in ways that underscore their persistence and mobility. These "forever chemicals" spread through multiple pathways, contaminate vital water resources, and accumulate in food chains, leading to wide-reaching consequences for ecological and human health. Understanding how PFAS move through and interact with the environment is critical for addressing their impacts and developing strategies to mitigate this global crisis. This chapter explores the pathways of contamination, the processes of bioaccumulation and biomagnification in the food chain, and the role of the global water cycle in the distribution of PFAS.

Pathways of PFAS Entering Surface and Groundwater

PFAS enter the environment through numerous pathways, including industrial discharges, municipal wastewater, landfill leachate, atmospheric deposition, and even human excretion. These pathways facilitate the widespread distribution of PFAS across surface and groundwater systems, soil, and even urban infrastructure like streets and pavements. Each pathway represents a critical source of contamination, emphasizing the complex and interconnected nature of PFAS pollution. The following are examples of sources and generation of PFAS to the wider environment:

Industrial Discharges

Industrial facilities that manufacture or use PFAS in their processes are among the most significant contributors to PFAS contamination. These facilities often release untreated waste directly into nearby water bodies, soil, or air. Common industries implicated include chemical manufacturing, electronics production, and textile treatment facilities that use PFAS for stain or water resistance. Once released into surface water, PFAS are carried downstream, spreading contamination across large geographic areas and affecting drinking water supplies, aquatic ecosystems, and recreational water bodies (Cousins et al., 2019).

In many cases, industrial sites also discharge PFAS-laden effluents into municipal wastewater systems, further exacerbating contamination. Industrial stormwater runoff, which collects PFAS residues from facility grounds, is another source of direct contamination to nearby rivers and lakes. The cumulative effects of these discharges create long-term environmental challenges, as PFAS persist indefinitely and continue to spread far from the original release points.

Municipal Wastewater and Human Excretion

Municipal wastewater is another major pathway for PFAS contamination, with human excretion playing a significant role. Once PFAS enter the human body through contaminated food, water, or products, they are not fully metabolized or broken down. Instead, they are excreted in urine and feces, where they re-enter municipal wastewater systems. A study by Schultz et al. (2006) highlighted the presence of PFAS in municipal sewage, with measurable concentrations of perfluoroalkyl acids (PFAAs) such as PFOA and PFOS in raw and treated wastewater. Wastewater treatment plants, however, are not designed to remove PFAS effectively. As a result,

PFAS are discharged into surface water or bound to biosolids, which are often applied as fertilizers to agricultural fields, spreading contamination to soils and crops (Nguyen et al., 2020).

Urban wastewater systems also collect runoff from streets and pavements, which can contain PFAS from various sources, such as fire-fighting foam residues, vehicle emissions, and leaching from treated materials like asphalt sealants. This runoff is often funneled into storm drains that lead directly to rivers and lakes, bypassing treatment entirely. Pavements themselves can act as sources of PFAS, as some sealants and coatings contain these chemicals to improve durability and water resistance. Over time, weathering and abrasion release PFAS into the environment, further contaminating urban and natural water systems (Wang et al., 2017).

Landfill Leachate

Landfills, often considered the final destination for consumer products, serve as long-term reservoirs for PFAS contamination. Products such as nonstick cookware, stain-resistant textiles, waterproof clothing, and food packaging frequently contain PFAS. When these products are disposed of in landfills, PFAS leach into surrounding soil and groundwater through landfill leachate. This leachate, a liquid formed by the decomposition of organic waste and the percolation of water through the landfill, often contains high concentrations of PFAS, especially PFOA and PFOS (Lang et al., 2017).

Efforts to treat landfill leachate are complicated by the persistence of PFAS, as conventional treatment methods such as biological processes, coagulation, or sedimentation are largely ineffective. Without advanced treatment technologies, PFAS from landfill leachate often migrate into nearby groundwater systems, creating a long-lasting source of contamination.

Streets, Pavements, and Urban Runoff

Urban environments contribute significantly to PFAS contamination through streets, pavements, and runoff. Pavement sealants, often used to extend the lifespan of asphalt surfaces, frequently contain PFAS to improve water resistance and durability. Over time, these chemicals leach into the environment due to weathering, abrasion, and exposure to sunlight. Rainfall and urban runoff then carry PFAS from streets and pavements into storm drains, which often discharge directly into rivers and lakes without treatment.

Additionally, firefighting activities in urban areas introduce PFAS to streets and pavements, particularly when aqueous film-forming foams (AFFFs) are used to combat fuel fires. Residues from these foams persist on surfaces and eventually wash into nearby water systems. This pathway underscores the need for stricter regulations and alternative materials that do not rely on PFAS.

Atmospheric Deposition and the Interplay of PFAS with Environmental Systems

Atmospheric deposition is one of the most pervasive and far-reaching pathways for PFAS contamination, enabling these chemicals to spread globally and affect regions far removed from industrial activity. Volatile PFAS precursors, such as fluorotelomer alcohols (FTOHs), are released into the air during the manufacturing of PFAS products, their use, or even the incineration of PFAS-containing waste. These precursors, due to their high volatility, are easily carried by air currents over long distances. As they move through the atmosphere, these compounds undergo chemical transformations, degrading into more stable perfluoroalkyl acids (PFAAs) such as perfluorooctanoic acid (PFOA) and perfluorooctanesulfonic acid (PFOS). Eventually, these stable compounds return to the Earth's surface through wet deposition (rain or snow) or dry deposition (settling of particles) (Young et al., 2007). This process contributes significantly to PFAS contamination in regions distant from their sources, including remote and pristine areas such as the Arctic and Antarctic. The detection of PFAS in polar regions underscores their global mobility, as well as the persistence and resilience of these "forever chemicals."

In urban areas, the effects of atmospheric deposition are equally concerning. Volatile PFAS precursors released from industrial emissions, consumer products, and waste incineration settle onto rooftops, streets, and pavements. Rainfall acts as a powerful vector for transporting these deposited PFAS into stormwater systems. From there, PFAS-contaminated runoffs are often funneled directly into surface water bodies, such as rivers, lakes, and reservoirs, bypassing any form of filtration or treatment. This pathway not only exacerbates the contamination of urban water systems but also introduces PFAS into broader aquatic ecosystems. The accumulation of PFAS on impervious urban surfaces, combined with their ability to leach into stormwater, highlights the critical role that atmospheric deposition plays in perpetuating urban water pollution.

Cyclical Nature of Atmospheric and Terrestrial PFAS Transport

A less-discussed yet equally important pathway is the interaction between atmospheric deposition and terrestrial systems, which creates a cyclical contamination process. Once volatile PFAS compounds are deposited onto land surfaces through precipitation, they interact with the soil and vegetation. In soils, PFAS exhibit complex behaviors. Some compounds bind tightly to organic matter and mineral surfaces, creating long-term reservoirs that can slowly release PFAS over time. Others leach into groundwater, contaminating aquifers and drinking water supplies. This infiltration process is particularly problematic in areas with high precipitation or porous soil, where deposited PFAS can travel deep into the subsurface, evading natural filtration mechanisms (Nguyen et al., 2020).

The interaction between atmospheric and terrestrial systems is further complicated by the potential for re-volatilization. Under certain environmental conditions, PFAS precursors deposited on land can evaporate back into the atmosphere, starting the cycle anew. This phenomenon, often referred to as the "grasshopper effect," enables PFAS to migrate across regions in a continuous cycle of deposition, evaporation, and re-deposition. This dynamic process ensures that PFAS are not confined to localized areas but instead perpetually circulate through interconnected environmental systems.

Complexity of Addressing Atmospheric PFAS Deposition

The interplay between atmospheric deposition and other environmental systems highlights the complexity of managing PFAS contamination. In rural and agricultural regions, precipitation containing PFAS can infiltrate soils and contaminate crops. In urban areas, the combination of atmospheric deposition and stormwater runoff introduces PFAS into municipal water supplies, making remediation efforts more challenging and costly. In remote regions, such as the Arctic and Antarctic, the deposition of PFAS exemplifies their ability to impact ecosystems with no direct industrial activity. For example, snowmelt in polar regions can release PFAS that were deposited over decades, reintroducing these chemicals into aquatic ecosystems where they bioaccumulate in wildlife such as polar bears and seals (Butt et al., 2010).

This cyclical transport of PFAS also complicates localized remediation efforts. Addressing PFAS contamination in one environmental compartment, such as soil or surface water, does not eliminate their presence in other interconnected systems (Butt et al., 2010; Lang et al., 2017; Young et al., 2007). For instance, efforts to reduce PFAS in urban water systems may be undermined by continued deposition from the atmosphere or leaching from contaminated soils. This interconnectivity underscores the need for comprehensive, system-wide approaches to managing PFAS pollution.

The atmospheric deposition of PFAS represents a critical pathway for their widespread contamination and underscores the interconnectedness of environmental systems. The ability of volatile PFAS precursors to travel long distances, settle on land and water surfaces, and interact with soil and groundwater demonstrates the global scale of the PFAS crisis. This cyclical movement between the atmosphere, terrestrial surfaces, and aquatic systems ensures that PFAS contamination persists and spreads, even in regions far removed from direct industrial sources. Addressing atmospheric deposition as part of broader PFAS mitigation strategies is essential, as localized interventions alone cannot disrupt the global transport and persistence of these chemicals. Only through coordinated efforts that span regulatory, technological, and environmental frameworks can we hope to break the cycle of PFAS contamination.

The pathways through which PFAS enter the environment are diverse and interconnected, highlighting their pervasive and persistent nature. Industrial discharges release PFAS directly into surface water and air, while municipal wastewater systems inadvertently reintroduce these chemicals into the environment through untreated effluents and the application of contaminated biosolids. Landfill leachate carries PFAS from disposed consumer products into soils and groundwater, further expanding contamination. Atmospheric deposition, however, represents one of the most far-reaching pathways. Volatile PFAS precursors released during manufacturing, product use, and waste incineration travel long distances via air currents, ultimately depositing onto land and water through precipitation or particle settling. This atmospheric transport enables PFAS to contaminate remote regions, including the Arctic and Antarctic, underscoring their global reach. Urban runoff from streets and pavements compounds this issue, as deposited PFAS are washed into stormwater systems and surface water bodies, often bypassing treatment. These interconnected pathways emphasize the cumulative impact of human activities, from industrial processes to everyday habits, perpetuating the presence of PFAS in the environment. Addressing PFAS contamination requires a comprehensive and systemic approach that considers all pathways, particularly the atmospheric dimension, to reduce emissions at every stage of the PFAS lifecycle. Without coordinated and far-reaching action, the environmental and health consequences of PFAS will continue to escalate, posing an increasingly complex challenge in managing these "forever chemicals."

Bioaccumulation and Biomagnification in the Food Chain

One of the most concerning aspects of PFAS is their ability to bioaccumulate and biomagnify in living organisms. Bioaccumulation occurs when organisms absorb PFAS from their environment, either through contaminated water, sediment, or food, and retain these chemicals in their tissues over time. Unlike many other contaminants, PFAS are not easily metabolized or excreted, which leads to their gradual buildup in the body. This persistence is particularly evident in long-lived species, such as fish, marine mammals, and birds, which can accumulate significant PFAS concentrations over their lifetimes (Domingo et al., 2021). For example, studies have found elevated PFAS levels in fish such as trout and salmon, which are often consumed by humans and wildlife (Ahrens & Bundschuh, 2014). This long-term retention has cascading effects on entire ecosystems, making bioaccumulation a central factor in PFAS contamination.

Biomagnification compounds the issue by increasing PFAS concentrations as they move up the food chain. For instance, small fish that have absorbed PFAS from contaminated water or prey are eaten by larger predators, which in turn are consumed by apex predators, including humans. With each trophic level, the PFAS concentration becomes more pronounced. This process is well-documented in aquatic ecosystems, where PFAS have been found in high concentrations in predatory fish such as tuna and swordfish, seabirds like gulls, and marine mammals such as seals and polar bears (Butt et al., 2010). In humans, consumption of seafood contaminated with PFAS has been linked to adverse health outcomes, including kidney and testicular cancers, immune system suppression, liver damage, and developmental issues in children (Grandjean & Clapp,

2015; Post et al., 2012). Furthermore, a recent study by Liu et al. (2021) highlighted the role of seafood as a major contributor to PFAS exposure in human populations, particularly in coastal and fish-reliant communities.

The persistence of PFAS in food chains poses significant ecological and economic risks. Ecologically, PFAS contamination can lead to declines in biodiversity by impacting the health and reproduction of exposed species. For instance, PFAS exposure in fish has been shown to disrupt endocrine functions, impair immune responses, and reduce growth rates (Nguyen et al., 2020). These effects ripple through aquatic ecosystems, destabilizing predator-prey relationships and altering the balance of populations. The contamination also extends to terrestrial ecosystems, as animals that feed on aquatic species, such as bears and birds of prey, accumulate PFAS from their diets.

Economically, the presence of PFAS in seafood has severe consequences for fisheries and aquaculture. Contaminated fish stocks may become unmarketable, leading to financial losses for fishers and disruptions to local economies that rely on these industries. In addition, regulatory restrictions on PFAS levels in food products have forced some fisheries to shut down entirely in areas of high contamination, further exacerbating economic impacts (Ahrens & Bundschuh, 2014). Public health costs related to PFAS exposure also place an additional economic burden on governments and healthcare systems, as communities near contaminated water bodies experience higher incidences of PFAS-related illnesses (Post et al., 2012).

The interconnected nature of food webs means that PFAS contamination in one area can have ripple effects throughout entire ecosystems, further emphasizing the urgency of comprehensive strategies to address these pollutants. These strategies must include stricter regulations on PFAS emissions, enhanced monitoring of contamination in water bodies and seafood, and investment in remediation technologies capable of reducing PFAS levels in ecosystems. Without intervention, the bioaccumulation and biomagnification of PFAS will continue to threaten biodiversity, public health, and global economies.

PFAS and the Global Water Cycle

The global water cycle plays a critical role in the distribution and persistence of PFAS, acting as a conveyor belt that transfers these chemicals between environmental compartments such as surface water, groundwater, soil, and air. PFAS are highly mobile due to their unique chemical properties, enabling them to spread widely across ecosystems. In surface water, PFAS remain dissolved for extended periods because of their hydrophilic nature, making them difficult to remove through natural processes. Additionally, their amphiphilic properties allow PFAS to bind to organic matter and sediments, creating long-term contamination reservoirs. These reservoirs can release PFAS back into the water column when physical disturbances, such as dredging or flooding, or chemical changes, such as shifts in pH, disrupt the equilibrium. This reintroduction of PFAS perpetuates contamination cycles and complicates remediation efforts (Nguyen et al., 2020; Cousins et al., 2019).

Groundwater systems are particularly vulnerable to PFAS contamination due to the chemicals' mobility through soil. PFAS can infiltrate groundwater from various sources, including landfill leachate, agricultural runoff, and firefighting foam applications. Once in groundwater, PFAS can travel long distances, often far from the original contamination site, due to their resistance to degradation and affinity for water. This persistence means that groundwater contamination can last for decades, posing significant challenges for drinking water safety. Groundwater serves as a primary source of drinking water for millions of people worldwide, and the presence of PFAS in these supplies has been linked to adverse health outcomes, including cancer and endocrine disruption (Post et al., 2012). Cleanup efforts for contaminated groundwater are both technically complex and costly, as traditional filtration and treatment systems are often insufficient to remove PFAS effectively (Lang et al., 2017).

Atmospheric processes further amplify the global spread of PFAS. Volatile PFAS precursors, such as fluorotelomer alcohols (FTOHs), are released into the atmosphere during manufacturing, product use, or waste incineration. These compounds can travel across continents via atmospheric currents before settling into surface water or soil through wet deposition (rain and snow) or dry deposition (settling of airborne particles). This atmospheric cycling explains why PFAS have been detected in remote regions like the Arctic and Antarctic, where there are no local sources of contamination. For example, studies have found significant PFAS concentrations in Arctic snow and ice, as well as in polar wildlife such as seals and polar bears, highlighting the global mobility and bioaccumulative potential of these chemicals (Young et al., 2007; Butt et al., 2010).

The water cycle perpetuates PFAS contamination through processes such as evaporation, condensation, and precipitation. PFAS can evaporate from contaminated water bodies, enter the atmosphere, and later return to the surface through rainfall. This continuous cycling ensures that PFAS remain a persistent and pervasive presence in the environment, even in regions far from their original source. For instance, PFAS-contaminated water from industrial sites can evaporate, only to return to agricultural fields or residential areas through precipitation, creating new pathways for human and ecological exposure (Grandjean & Clapp, 2015; Wang et al., 2017). This interconnectedness of environmental systems underscores the difficulty of containing PFAS contamination, as efforts to address one pathway can be undermined by reintroduction through another.

The persistent cycling of PFAS within the global water cycle highlights the need for coordinated international efforts to address their contamination. This includes stricter regulation of PFAS emissions, improved waste management practices, and the development of advanced remediation technologies capable of breaking down these "forever chemicals." Without a comprehensive approach that accounts for the role of the water cycle, PFAS contamination will continue to spread, posing ongoing risks to ecosystems, public health, and water resources.

Role of Sediments in PFAS Persistence

An often overlooked yet critical aspect of PFAS contamination is the role sediments play in acting as reservoirs for these persistent chemicals. In aquatic environments, PFAS readily bind to organic matter and fine sediments due to their amphiphilic nature, which allows them to interact with both hydrophobic and hydrophilic substances. Once bound, PFAS can remain sequestered in sediment layers for years or even decades. This binding process occurs because many PFAS compounds, particularly long-chain variants such as perfluorooctanoic acid (PFOA) and perfluorooctanesulfonic acid (PFOS), have a high affinity for organic carbon and other sediment components (Ahrens & Bundschuh, 2014; Cousins et al., 2019).

Sediments, therefore, act as long-term sources of PFAS contamination, gradually releasing these chemicals back into the water column under certain environmental conditions. Events such as flooding, storm surges, or dredging can physically disturb sediments, resuspending PFAS into the overlying water. Additionally, chemical changes in the aquatic environment, such as shifts in pH, redox potential, or salinity, can disrupt the bonds between PFAS and sediment particles, leading to their remobilization (Nguyen et al., 2020). For instance, studies have shown that dredging activities in contaminated rivers and lakes can significantly increase PFAS concentrations in surface water, highlighting the potential risks associated with sediment disturbance (Lang et al., 2017).

The persistent nature of PFAS in sediments not only complicates remediation efforts but also poses a continuous threat to aquatic ecosystems. Sediments serve as a primary habitat for benthic organisms, which are often at the base of aquatic food webs. These organisms can absorb PFAS from sediments, introducing the chemicals into the food chain and facilitating bioaccumulation and biomagnification (Domingo et al., 2021). This process can affect higher trophic levels, including predatory fish, birds, and mammals, as well as humans who consume contaminated seafood (Butt et al., 2010).

Moreover, the interactions between PFAS and sediments are influenced by site-specific factors, such as sediment composition, hydrology, and the type and concentration of PFAS present. For example, sediments with high organic carbon content are more likely to sequester PFAS, while fast-moving water bodies may transport PFAS-contaminated sediments to new locations, spreading contamination over a broader area (Grandjean & Clapp, 2015). Understanding these interactions is essential for developing effective remediation strategies. Current sediment remediation techniques, such as capping or dredging, are often inadequate for addressing PFAS contamination because they fail to break down the chemicals or prevent their re-release into the environment (Nguyen et al., 2020).

Advanced remediation methods are being explored to address the challenges posed by PFAS-contaminated sediments. Technologies such as thermal desorption, chemical oxidation, and in situ stabilization show promise, but they are often costly and technically complex, limiting their widespread application (Wang et al., 2017). In the absence of effective large-scale solutions, managing PFAS in sediments remains a critical challenge for environmental scientists, policymakers, and industries.

Understanding the role of sediments as PFAS reservoirs is crucial for developing holistic management strategies. Addressing sediment contamination requires integrated approaches that consider the interactions between sediments, water, and biota. Without targeted interventions, sediments will continue to act as sources of PFAS, perpetuating their presence in aquatic ecosystems and compounding the risks to environmental and human health.

Ultimately, PFAS contamination in the wider environment represents a daunting legacy of modern industrial progress; one that intertwines environmental, economic, and public health challenges on a global scale. The persistence of these "forever chemicals" in surface and groundwater, their insidious accumulation in food chains, and their relentless global spread through the water cycle highlight the systemic nature of this crisis. This is not merely an issue of local pollution but a pervasive threat that transcends borders, ecosystems, and generations. Addressing PFAS contamination demands more than isolated interventions; it requires an unwavering commitment to understanding their environmental behavior, fostering international collaboration, and prioritizing innovative solutions that can break the cycle of contamination. As PFAS continue to circulate through air, water, and soil, we stand at a critical juncture; where the urgency to act decisively is matched only by the opportunity to redefine how we confront environmental challenges. The path forward must be one of innovation, regulation, and shared responsibility to protect ecosystems and human health for generations to come.

Chapter 5

Global Scope of PFAS Contamination

PFAS contamination is not just a localized issue confined to industrial areas or specific nations—it is a global crisis affecting nearly every corner of the planet. From heavily industrialized regions to the most remote ecosystems, PFAS have been detected in water, soil, and even air. These "forever chemicals" persist across continents, infiltrating rivers, lakes, and groundwater, while their presence highlights gaps in regulation and enforcement worldwide. Understanding the global scope of PFAS contamination requires examining major hotspots, specific case studies from diverse geographic regions, and the regulatory disparities that hinder comprehensive mitigation efforts. Only by addressing PFAS as a global problem can we begin to develop effective, unified strategies to combat their pervasive impacts.

Major Hotspots of Contamination Worldwide

PFAS contamination is most pronounced in regions with a history of industrial activity, military installations, and heavy reliance on PFAS-containing products. The United States, for example, has identified thousands of contaminated sites, including areas near military bases where firefighting foams containing PFAS were used extensively. In

Europe, hotspots include regions with fluorochemical manufacturing facilities, such as Belgium's Flanders region and parts of Germany (Grandjean & Clapp, 2015). Similar contamination has been documented in Japan, where PFAS concentrations have been linked to industrial discharges and municipal wastewater (Nguyen et al., 2020).

Emerging economies, including nations in Asia, Africa, and South America, are also grappling with PFAS contamination, often without the regulatory frameworks or detection capabilities to address it effectively. For example, high levels of PFAS have been reported in China's Pearl River Delta, a heavily industrialized region where PFAS contamination has spread through surface water, groundwater, and sediment (Liu et al., 2021). In Africa, limited monitoring has revealed PFAS in urban rivers and drinking water, indicating potential widespread contamination that remains underreported (Ahrens & Bundschuh, 2014). These hotspots illustrate how PFAS contamination transcends borders, creating shared challenges for nations worldwide.

Case Studies: Rivers and Lakes

Europe: The Rhine River

The Rhine River, one of Europe's most significant waterways, flows through six countries, Switzerland, Liechtenstein, Austria, Germany, France, and the Netherlands, before emptying into the North Sea. As one of the most industrialized river basins in the world, the Rhine has become a major conduit for PFAS contamination. Studies have consistently documented elevated PFAS levels in its waters, largely attributed to industrial discharges, municipal wastewater treatment plant effluents, and stormwater runoff from urban areas. Industrial hubs along the Rhine, particularly in Germany and Switzerland, are notable sources of PFAS, with chemicals entering the river system from fluorochemical manufacturing facilities, chemical plants, and textile industries (Cousins et al., 2019).

One major challenge in managing PFAS contamination in the Rhine is its transboundary nature. As the river flows through multiple nations, contaminants introduced upstream can easily affect downstream regions, complicating efforts to regulate and remediate PFAS inputs. For example, industries in Switzerland and Germany contribute significantly to PFAS loads that are ultimately detected in the Netherlands, where the river discharges into the North Sea. This highlights the necessity of collaborative efforts between nations to monitor and reduce PFAS contamination.

Sources of Contamination

A significant proportion of PFAS in the Rhine comes from wastewater treatment plants (WWTPs), which are not equipped to remove PFAS effectively. These facilities serve highly industrialized areas and discharge treated wastewater containing PFAS directly into the river. Additionally, stormwater runoff from urban centers contributes to PFAS inputs, particularly during heavy rainfall events that wash contaminants from roads, rooftops, and paved surfaces into the river. Agricultural runoff from areas near the Rhine also carries PFAS from biosolids and fertilizers applied to fields (Ahrens & Bundschuh, 2014).

Industrial hotspots along the Rhine exacerbate the contamination problem. For instance, regions in Switzerland and Germany with chemical and fluorochemical manufacturing facilities have been identified as key contributors to PFAS levels in the river. In some cases, historical contamination from legacy PFAS compounds, such as PFOA and PFOS, continues to leach into the river from sediments and surrounding soils.

Monitoring and Remediation Efforts

Germany, Switzerland, and the Netherlands have implemented robust monitoring programs to track PFAS concentrations in the Rhine and its tributaries. These programs involve regular sampling of surface water, sediments, and biota to assess contamination levels and trends. The International Commission for the Protection of the Rhine (ICPR), a transboundary organization, plays a central role in coordinating monitoring and remediation efforts among the riparian nations. However, despite these efforts, reducing PFAS inputs remains a significant challenge due to the diffuse and persistent nature of the contamination (Nguyen et al., 2020).

Remediation efforts in the Rhine are complicated by the chemical stability of PFAS and the diverse sources of contamination. Traditional water treatment technologies, such as filtration and sedimentation, are largely ineffective at removing PFAS, necessitating the development and implementation of advanced treatment methods like activated carbon filtration and reverse osmosis. However, these methods are costly and not feasible for large-scale application across the entire river system. Moreover, contaminated sediments in the Rhine act as long-term reservoirs for PFAS, releasing the chemicals back into the water column during events such as flooding or dredging (Lang et al., 2017).

Cross-Border Collaboration and Policy

The transboundary nature of the Rhine requires strong cross-border collaboration to address PFAS contamination effectively. The ICPR has been instrumental in fostering cooperation among riparian countries, promoting information sharing, and developing joint action plans. However, disparities in regulatory frameworks and enforcement across the Rhine basin pose significant hurdles. While the European Union has introduced stringent regulations on certain PFAS compounds, such as PFOS, individual nations along the Rhine vary in their implementation and enforcement of these measures (Grandjean & Clapp, 2015).

Efforts to remediate the Rhine highlight the complexities of managing PFAS contamination in transboundary water systems. As upstream activities directly impact downstream water quality, addressing PFAS in the Rhine requires not only technological advancements but also coordinated policy and regulatory approaches among the affected nations. This case underscores the importance of international cooperation in tackling the global challenge of PFAS contamination.

Asia: The Ganges River

The Ganges River, revered as a sacred waterway in India, is also one of the most polluted river systems globally, and PFAS contamination adds a dangerous dimension to its environmental challenges. The Ganges basin, home to over 400 million people, serves as a crucial water source for drinking, agriculture, and industry. However, the presence of PFAS compounds in its waters highlights the intersection of unchecked industrialization, inadequate wastewater management, and environmental neglect. Studies have consistently documented alarming PFAS levels in the Ganges, which stem from untreated industrial effluents, urban runoff, and agricultural wastewater (Nguyen et al., 2020; Cousins et al., 2019).

Sources of PFAS Contamination

Industrial discharges are the most prominent source of PFAS contamination in the Ganges River. India is a growing hub for chemical manufacturing, including textiles, leather processing, and paper production, industries known to utilize PFAS in their processes. Effluents from these industries are often discharged directly into the river without proper treatment, significantly contributing to PFAS levels in the water. For example, Kanpur, a major industrial city along the Ganges, is a hotspot for leather tanneries that release untreated waste containing PFAS and other hazardous chemicals (Ahrens & Bundschuh, 2014).

Urban runoff also plays a critical role in PFAS contamination. Cities along the river, such as Varanasi and Kolkata, generate significant amounts of runoff containing PFAS from household products, firefighting foams, and untreated domestic wastewater. Rainwater washes these chemicals into the river, compounding contamination levels, especially during the monsoon season. Additionally, poorly managed landfills in urban areas leach PFAS into nearby water bodies and groundwater, further exacerbating the issue.

Agricultural activities along the Ganges basin also contribute to PFAS contamination. Farmers frequently use biosolids and sludge from wastewater treatment plants as fertilizers, inadvertently introducing PFAS into agricultural runoff. This runoff, laden with PFAS, enters the river and nearby tributaries, creating a feedback loop that sustains contamination levels (Grandjean & Clapp, 2015).

Impacts on Drinking Water and Public Health

The contamination of the Ganges River by PFAS poses serious risks to millions of residents who rely on it for drinking water. Studies have detected PFAS compounds, such as perfluorooctanoic acid (PFOA) and perfluorooctanesulfonic acid (PFOS), in municipal water supplies derived from the river. Even at low concentrations, long-term exposure to PFAS is associated with adverse health outcomes, including kidney and liver damage, immune suppression, and developmental issues in children (Domingo et al., 2021; Post et al., 2012).

Communities living along the Ganges often lack access to advanced water treatment facilities capable of removing PFAS, leaving them vulnerable to prolonged exposure. Rural areas, in particular, face significant challenges, as many residents rely on untreated water directly from the river for daily use. This exposure exacerbates public health disparities and places an additional burden on already strained healthcare systems.

Addressing PFAS in the Ganges River

Efforts to mitigate PFAS contamination in the Ganges require a multifaceted approach. Stricter industrial regulations are essential to control the release of PFAS from manufacturing facilities. Regulatory agencies need to enforce stricter effluent discharge standards and ensure that industries adopt advanced wastewater treatment technologies capable of removing PFAS (Nguyen et al., 2020).

Investing in modern water treatment infrastructure is equally critical. Traditional wastewater treatment plants are ineffective at removing PFAS, necessitating the implementation of advanced methods such as activated carbon filtration, reverse osmosis, and ion exchange. However, these technologies are costly and require significant government investment and international support to be scaled across the basin.

Community awareness and education also play a vital role in addressing PFAS contamination. Informing local populations about the risks of PFAS and promoting safe water practices can help reduce exposure. International collaboration, including technology transfer and funding from global environmental organizations, is necessary to support India in tackling this complex issue.

The Ganges River exemplifies the challenges of managing PFAS contamination in a highly populated and industrialized region. The combination of industrial discharges, urban runoff, and agricultural wastewater underscores the systemic nature of the problem. Addressing PFAS in the Ganges requires a comprehensive strategy involving stricter regulations, advanced water treatment technologies, and international cooperation. Without urgent action, the health risks and ecological damage caused by PFAS will continue to escalate, threatening millions of lives and one of the world's most iconic river systems.

Africa: The Umgeni River

The Umgeni River, flowing through the KwaZulu-Natal region in South Africa, is a critical lifeline for millions of people. It provides water for drinking, agriculture, and industrial activities, sustaining both urban centers like Durban and rural communities in the region.

However, the river faces growing threats from contamination, including alarming levels of per- and polyfluoroalkyl substances (PFAS). PFAS contamination in the Umgeni River primarily stems from industrial effluents, urban wastewater, and stormwater runoff. This contamination has significant implications for human health, ecosystems, and water security in the region (Domingo et al., 2021; Ahrens & Bundschuh, 2014).

Sources of PFAS Contamination

One of the primary contributors to PFAS contamination in the Umgeni River is industrial discharges. The KwaZulu-Natal region is home to numerous industries, including textiles, chemicals, and manufacturing, many of which rely on PFAS for their water- and stain-resistant properties. Effluents from these industries are often discharged directly into the river or its tributaries without adequate treatment, introducing high concentrations of PFAS into the water system (Nguyen et al., 2020).

Urban wastewater further exacerbates the contamination. Cities and towns along the Umgeni River generate significant amounts of domestic and industrial wastewater, much of which is processed by outdated wastewater treatment plants. These facilities are ill-equipped to filter out PFAS, allowing the chemicals to pass through and re-enter the river system. Additionally, informal settlements lacking proper sanitation infrastructure contribute to PFAS levels through untreated sewage and runoff from consumer products containing PFAS (Grandjean & Clapp, 2015).

Stormwater runoff also plays a critical role in PFAS contamination. During heavy rainfall, water washes contaminants from roads, rooftops, and industrial sites into the river. These contaminants include PFAS from firefighting foams, household products, and industrial activities. The seasonal nature of this runoff means PFAS levels can spike during periods of heavy rain, such as South Africa's summer rainy season (Cousins et al., 2019).

Impacts on Drinking Water and Ecosystems

The Umgeni River is a vital source of drinking water for the KwaZulu-Natal region, serving millions of residents, including the major metropolitan area of Durban. PFAS contamination in this water poses significant health risks, particularly for vulnerable populations in rural and urban areas who depend directly on untreated or minimally treated water from the river. PFAS exposure is linked to adverse health effects such as cancer, immune system suppression, and developmental issues in children, raising concerns about long-term public health outcomes (Post et al., 2012; Domingo et al., 2021).

Ecologically, the presence of PFAS in the Umgeni River disrupts aquatic ecosystems. PFAS accumulate in sediments, which act as long-term reservoirs for the chemicals. Bottom-dwelling organisms absorb PFAS from the sediments, introducing the chemicals into the food chain. Predatory fish, birds, and other wildlife that rely on the river for sustenance face increased risks of bioaccumulation and biomagnification, threatening biodiversity and ecosystem stability (Butt et al., 2010).

Challenges in Addressing PFAS Contamination

Efforts to address PFAS contamination in the Umgeni River are hindered by several challenges. South Africa's wastewater treatment infrastructure is outdated and underfunded, making it difficult to effectively remove PFAS from industrial and municipal wastewater. Furthermore, enforcement of environmental regulations is inconsistent, with limited oversight of industries discharging into the river. These gaps in governance allow PFAS contamination to persist and spread unchecked (Nguyen et al., 2020).

Monitoring and remediation efforts are also limited by resource constraints. While PFAS detection requires sophisticated analytical techniques like high-resolution mass spectrometry, these technologies are not widely available in developing regions like South Africa. This lack of monitoring capacity means that contamination often goes undetected until it reaches critical levels, further delaying response efforts. Additionally, the high cost of advanced water treatment technologies, such as reverse osmosis and activated carbon filtration, makes widespread implementation challenging for regions with limited financial resources (Grandjean & Clapp, 2015).

Need for International Support

Addressing PFAS contamination in the Umgeni River will require international support and collaboration. Technology transfer from developed nations could help South Africa acquire the advanced analytical and remediation tools needed to tackle PFAS effectively. Funding from global environmental organizations and international development agencies could support the modernization of wastewater treatment infrastructure and the implementation of stricter industrial regulations. Community engagement and education campaigns are also critical for raising awareness about PFAS risks and promoting safe water practices among local populations (Domingo et al., 2021).

The PFAS contamination of the Umgeni River underscores the complex challenges faced by developing regions in managing industrial pollutants. As a vital water source for millions, the river's contamination threatens public health, biodiversity, and water security. Addressing this issue will require a coordinated approach involving stricter regulations, improved wastewater treatment, and international collaboration. Without immediate action, PFAS contamination in the Umgeni River will continue to escalate, jeopardizing the well-being of communities and ecosystems alike.

North America: Rivers of the Western United States

The Western United States is home to some of the most vital river systems that sustain both ecosystems and human populations. However, these rivers are increasingly burdened by the presence of per- and polyfluoroalkyl substances (PFAS), a class of chemicals known for their persistence, mobility, and toxicity. A study by Sims et al. (2025) provided critical insights into the prevalence and ecological impact of PFAS across 16 major rivers in this region, illustrating the global scope and interconnected nature of PFAS contamination.

Sampling and Analytical Methods

The study involved comprehensive sampling from October 2022 to March 2023, targeting rivers across ten western states (Sims et al., 2025). Researchers employed Liquid Chromatography-Tandem Mass Spectrometry (LC-MS/MS) to identify and quantify seven key PFAS compounds, including perfluorooctanoate (PFOA) and perfluoroheptanoic acid (PFHpA). Sampling occurred upstream and downstream of wastewater treatment plants (WWTPs), emphasizing the role of anthropogenic activities in PFAS dissemination.

Findings: Contamination Hotspots and Patterns

The study revealed significant variability in PFAS concentrations among rivers. The Gila, Los Angeles, San Juan, and Santa Cruz Rivers exhibited the highest levels of PFOA, consistent with industrial and urban influences in their watersheds. Conversely, rivers like the Boise, Columbia, and Willamette displayed lower PFAS levels, underscoring regional disparities in contamination sources. PFHpA was particularly concerning, with concentrations in the Gila River reaching 199 ng/L, highlighting the compound's growing prevalence due to its designation as a "safer alternative" to legacy PFAS.

Source Attribution and Environmental Pathways

Principal Component Analysis (PCA) identified fluorotelomer alcohol degradation as a significant source of PFAS in rivers like the Boise and Clark Fork. WWTP discharges, atmospheric deposition, and industrial activities were pinpointed as primary contributors to contamination. The study emphasized that even remote rivers are not immune, as PFAS compounds are transported via atmospheric and hydrological pathways.

Ecological Impacts and Risk Assessment

Using hazard quotient (HQ) calculations, the study assessed the risks posed by PFAS to aquatic organisms like *Daphnia magna*. Most PFAS exhibited HQ values below levels of concern, yet certain compounds, such as PFDA in the Santa Cruz River, posed low but notable risks. The findings underscore the need for ongoing monitoring and site-specific risk assessments to address potential long-term ecological impacts.

Implications for Policy and Remediation

The results of this study have profound implications for environmental policy and remediation strategies. The pervasive presence of PFAS in Western U.S. rivers reflects broader global patterns of contamination, requiring coordinated regulatory frameworks and innovative cleanup technologies. As Sims et al. (2025) noted, PFAS contamination is not merely a regional issue but a global crisis that demands immediate and sustained action.

This case study serves as a critical reminder of the complexity of PFAS contamination and the urgent need for multidisciplinary approaches to mitigate its far-reaching effects.

South America: Lake Titicaca

Lake Titicaca, situated between Bolivia and Peru, is the highest navigable lake in the world and holds immense cultural, ecological, and economic significance for the Andean region. Despite its remote and seemingly pristine location, Lake Titicaca has not been immune to the global issue of PFAS contamination. Recent studies have detected PFAS compounds in its waters, primarily attributed to urban runoff, agricultural activities, and untreated wastewater from nearby cities and towns. This case highlights how PFAS can infiltrate even the most isolated ecosystems, posing severe risks to biodiversity, public health, and the livelihoods of local communities (Butt et al., 2010; Domingo et al., 2021).

Sources of PFAS Contamination

One of the primary sources of PFAS in Lake Titicaca is urban runoff. The rapidly growing populations of cities near the lake, such as Puno in Peru and Copacabana in Bolivia, contribute significantly to PFAS contamination. Urban runoff carries a mix of contaminants, including PFAS from consumer products such as nonstick cookware, water-repellent textiles, and food packaging. The absence of adequate stormwater management systems exacerbates this issue, allowing untreated runoff to flow directly into the lake (Nguyen et al., 2020).

Agricultural activities in the surrounding region further contribute to PFAS contamination. Farmers often use biosolids and sludge from wastewater treatment plants as fertilizers, which can contain significant levels of PFAS. These chemicals are then carried into the lake through agricultural runoff, particularly during heavy rainfall. Additionally, firefighting foams containing PFAS, used during agricultural burns or industrial incidents, may also leach into the lake over time, adding to its contamination levels (Cousins et al., 2019).

The lack of effective wastewater treatment infrastructure is another major contributor. Many communities around Lake Titicaca discharge untreated or minimally treated wastewater directly into the lake. This wastewater contains PFAS from both domestic and industrial sources, including detergents, cleaning agents, and industrial effluents. The combined impact of urban runoff, agricultural practices, and untreated wastewater creates a complex contamination matrix that is difficult to address (Ahrens & Bundschuh, 2014).

Impacts on Ecosystems and Communities

The presence of PFAS in Lake Titicaca has far-reaching consequences for its ecosystems and the communities that depend on it. Aquatic species, such as native fish and amphibians, are at risk due to PFAS bioaccumulation. These chemicals can accumulate in sediments and enter the food chain, disrupting aquatic ecosystems and threatening biodiversity. For example, local fish species such as *Orestias* are vital for both ecological balance and the livelihoods of fishing communities. PFAS contamination poses a dual threat, reducing fish populations and raising concerns about the safety of consuming contaminated fish (Butt et al., 2010).

For local communities, Lake Titicaca is a primary source of drinking water, fishing, and agriculture. PFAS contamination in the lake jeopardizes the health of residents who rely on untreated or minimally treated water for daily use. Long-term exposure to PFAS has been linked to serious health issues, including cancer, immune system suppression, and developmental problems in children. Moreover, the contamination threatens the economic stability of fishing and agricultural sectors that are integral to the region's livelihood and cultural identity (Domingo et al., 2021).

Challenges in Addressing PFAS Contamination

Efforts to address PFAS contamination in Lake Titicaca face significant challenges, particularly due to resource limitations in the region. Both Bolivia and Peru lack the advanced wastewater treatment infrastructure necessary to remove PFAS effectively from municipal and industrial effluents. Existing treatment plants are not equipped to handle the chemical stability of PFAS, allowing these compounds to persist in the lake's waters and sediments (Nguyen et al., 2020).

Monitoring PFAS levels in Lake Titicaca is another obstacle. High-resolution mass spectrometry, the gold standard for PFAS detection, is not readily available in the region. As a result, contamination often goes undetected or is underreported, delaying mitigation efforts. Furthermore, enforcement of environmental regulations is inconsistent, with limited resources and competing economic priorities often taking precedence over environmental protection (Grandjean & Clapp, 2015).

Need for Regional and International Partnerships

Addressing PFAS contamination in Lake Titicaca will require coordinated efforts at local, national, and international levels. Investments in advanced wastewater treatment infrastructure are critical to reducing PFAS inputs from urban and industrial sources. Funding and technical support from international organizations, such as the United Nations or the Global Environment Facility, could help both Peru and Bolivia build the capacity needed to tackle this issue effectively.

Community engagement and education campaigns are also essential. Informing local populations about the risks of PFAS and promoting safe water practices can reduce exposure and encourage public support for remediation efforts. Additionally, cross-border collaboration between Bolivia and Peru is crucial to ensure a unified approach to managing PFAS contamination in the shared waterway.

The case of Lake Titicaca underscores the pervasive nature of PFAS contamination and its ability to infiltrate even remote ecosystems. The combination of urban runoff, agricultural activities, and untreated wastewater presents a complex challenge that threatens both biodiversity and the livelihoods of local communities. Addressing PFAS in Lake Titicaca will require a multi-faceted strategy involving infrastructure improvements, regulatory enforcement, and international collaboration. Without urgent action, the lake's ecological and cultural significance may be irreparably compromised, leaving future generations to contend with the legacy of these "forever chemicals."

Global Disparities in Managing PFAS Contamination

Efforts to manage PFAS contamination on a global scale are significantly hindered by disparities in regulatory frameworks, detection capabilities, and enforcement mechanisms. While developed nations have made notable progress in establishing regulations and monitoring systems, many developing countries struggle with resource limitations and lack of infrastructure, leaving vulnerable populations exposed to the risks of PFAS contamination. This inequity underscores the need for global cooperation and targeted initiatives to address the widespread environmental and health impacts of these chemicals.

Regulations in Developed Nations

Developed nations, including the United States, European Union (EU) member states, and Japan, have implemented advanced regulatory frameworks to manage PFAS contamination. In the U.S., the Environmental Protection Agency (EPA) has issued health advisories for PFOA and PFOS in drinking water, setting limits as low as 0.004 parts per trillion (ppt) and 0.02 ppt, respectively, to protect public health (U.S. EPA, 2022). These advisories are part of broader regulatory efforts to phase out the use of PFAS in manufacturing and encourage the development of safer alternatives.

Similarly, the European Chemicals Agency (ECHA) has proposed banning PFAS in nonessential applications under the European Union's REACH (Registration, Evaluation, Authorization, and Restriction of Chemicals) framework. This ambitious initiative aims to significantly reduce the release of PFAS into the environment and limit human exposure. In Japan, strict regulations on PFOS and PFOA have been in place since 2010, with monitoring programs focusing on industrial hotspots and contaminated water sources (Nguyen et al., 2020).

Despite these advances, challenges persist even in developed nations. Enforcement of regulations varies between jurisdictions, and legacy PFAS contamination from past industrial activities continues to pose risks to water supplies and ecosystems. Moreover, addressing contamination at sites with historical PFAS use remains a costly and complex undertaking, requiring advanced remediation technologies and long-term management strategies (Grandjean & Clapp, 2015).

Challenges in Developing Nations

In contrast, many developing nations lack the resources to implement comprehensive PFAS monitoring and management programs. The high costs associated with advanced detection technologies, such as high-resolution mass spectrometry, make it difficult for these countries to accurately assess contamination levels. Instead, less sensitive and outdated techniques are often used, leading to underreporting of contamination and insufficient data for informed decision-making (Liu et al., 2021).

Regulatory gaps further exacerbate the issue. Developing nations often prioritize economic growth over environmental protection, resulting in weak enforcement of existing environmental laws or a complete absence of PFAS-specific regulations. For instance, industrial discharges, untreated wastewater, and landfill leachate containing PFAS frequently enter water systems unchecked, putting millions at risk of exposure. Vulnerable populations in rural areas and informal settlements are particularly affected, as they often rely on untreated water sources for drinking, agriculture, and daily use (Domingo et al., 2021).

International Initiatives and Global Cooperation

Efforts to bridge these disparities include international initiatives aimed at promoting collaboration, funding, and technology transfer. The Stockholm Convention on Persistent Organic Pollutants, an international treaty, has added certain PFAS compounds, such as PFOS, to its list of restricted chemicals. While this marks an important step in regulating PFAS globally, enforcement remains inconsistent due to differing levels of commitment and capacity among member nations (Grandjean & Clapp, 2015).

Global organizations, such as the United Nations Environment Program (UNEP) and the Global Environment Facility (GEF), play a critical role in supporting developing nations. These organizations provide funding and technical assistance for building capacity in PFAS detection, monitoring, and remediation. For example, UNEP has launched initiatives to train local scientists in PFAS sampling and analysis, enabling more accurate assessment of contamination in underserved regions (Nguyen et al., 2020).

Technology transfer from developed to developing nations is another key strategy. Sharing advanced analytical techniques, such as liquid chromatography-tandem mass spectrometry (LC-MS/MS), and remediation technologies, such as activated carbon filtration and reverse osmosis, can help developing nations address PFAS contamination more effectively. Capacity-building programs that train local officials and scientists in the use of these technologies are essential to ensuring their sustainable application (Liu et al., 2021).

The Path Forward

Addressing the global challenges of PFAS contamination requires coordinated efforts among nations to overcome regulatory, technical, and financial barriers. Developed nations must take the lead in providing funding, sharing technology, and supporting the capacity-building efforts of developing countries. International treaties like the Stockholm Convention need to be strengthened to ensure consistent enforcement and broader inclusion of PFAS compounds. Furthermore, community engagement and education campaigns are critical to raising awareness of PFAS risks and promoting safe practices in affected areas.

The disparities in PFAS management not only highlight the uneven distribution of resources and capabilities but also underscore the interconnected nature of environmental challenges. As PFAS contamination continues to spread through air, water, and soil, global cooperation is not just desirable, it is essential. Only through a unified approach can the world effectively tackle the pervasive threat of these "forever chemicals" and protect ecosystems and human health for future generations.

Finally, the global scope of PFAS contamination underscores a profound and interconnected environmental crisis that knows no borders. From the heavily industrialized rivers of Europe to the sacred waters of the Ganges, the high-altitude Lake Titicaca, and the vital Umgeni River in Africa, PFAS have infiltrated ecosystems and communities on every continent, threatening biodiversity, public health, and water security. These "forever chemicals" have exposed the glaring disparities in regulatory frameworks, detection capabilities, and remediation resources, emphasizing the urgent need for a unified international response.

As the world confronts this pervasive pollution, the path forward demands bold action: nations must collaborate to establish consistent regulatory standards, invest in cutting-edge technologies for detection and remediation, and ensure equitable access to resources for both developed and developing regions. Addressing the global challenge of PFAS contamination is not merely a scientific or political endeavor— it is a moral imperative to protect our planet and its inhabitants. Only through coordinated and sustained efforts can we begin to reverse the damage caused by these persistent pollutants and safeguard a sustainable future for generations to come.

Chapter 6

PFAS in the United States

The United States faces one of the most pressing challenges in addressing PFAS contamination due to the widespread presence of these chemicals in surface waters, drinking water supplies, and the broader environment. From industrial discharges to military installations, the pathways through which PFAS enter ecosystems are diverse and pervasive. This chapter delves into the scale of PFAS contamination in the United States, exploring case studies of major water systems and examining the significant roles played by military bases, manufacturing facilities, and landfills in contributing to the problem.

Overview of Contamination in U.S. Surface Waters

PFAS contamination in the United States is extensive, with traces of these chemicals detected in nearly every state. According to a 2022 report by the Environmental Working Group (EWG), more than 2,800 locations across the country have documented PFAS contamination in water supplies. These locations include rivers, lakes, and streams that serve as primary sources of drinking water for millions of Americans (EWG, 2022). PFAS pollution in these water bodies poses a serious threat to ecosystems and public health, with recent studies highlighting increasing concentrations in areas surrounding industrial facilities, military bases, and urban centers (Cousins et al., 2019).

One of the primary reasons for the widespread contamination is the historical use of PFAS in industrial and military applications. For decades, manufacturers released untreated PFAS waste into surface waters, leading to contamination in major river systems, including the Mississippi and the Ohio Rivers. Industrial hubs such as those in Michigan and West Virginia have reported some of the highest PFAS levels in the country due to prolonged discharge from chemical plants (Bilott, 2019). Additionally, military installations, which used firefighting foams containing PFAS for training and emergency preparedness, have emerged as significant contributors to contamination. These activities, often conducted near aquifers or surface water bodies, have introduced PFAS into critical drinking water sources, with some regions reporting levels far exceeding health advisory limits (Grandjean & Clapp, 2015).

Municipal wastewater treatment plants also play a role in PFAS contamination. These facilities, which are not designed to filter or degrade PFAS, inadvertently reintroduce these chemicals into the environment. Treated wastewater discharged into rivers and lakes often contains measurable PFAS concentrations, while biosolids applied to agricultural fields such as fertilizer can leach PFAS into soil and groundwater. This cyclical contamination underscores the challenges of containing PFAS pollution in urbanized regions and highlights the need for improved wastewater treatment technologies to address these persistent pollutants (Nguyen et al., 2020; Schultz et al., 2006).

The cumulative impacts of these contamination sources are particularly evident in areas where multiple pathways overlap. For instance, regions near industrial zones, military bases, and landfills often experience higher levels of PFAS in both surface and groundwater. These findings illustrate the interconnected nature of

PFAS contamination in the United States and the urgent need for comprehensive strategies to mitigate their spread and protect water resources.

Case Studies from the United States

Mississippi River

The Mississippi River, the largest river system in the United States, plays a critical role in supporting agriculture, industry, and drinking water supplies for millions of Americans. However, PFAS contamination has emerged as a growing concern along its extensive network. Industrial facilities located along the river's banks, particularly in the Midwest and Gulf regions, have been significant contributors to PFAS pollution. These facilities release PFAS-containing effluents into the river, often as a byproduct of manufacturing processes involving chemicals, textiles, and nonstick materials. Additionally, stormwater runoff from these industrial zones exacerbates the contamination, transporting PFAS into the waterway during heavy rains or flooding events (Schultz et al., 2006).

Municipal wastewater treatment plants also play a role in PFAS pollution in the Mississippi River. These plants are not designed to effectively remove PFAS, allowing these "forever chemicals" to pass through treatment systems and re-enter the river. Agricultural activities further compound the issue, as PFAS-laden biosolids are often applied as fertilizers, leading to runoff during irrigation or rainfall. The cumulative effect of these sources has resulted in detectable PFAS concentrations not only in the river itself but also in sediments and aquatic life.

The impacts of PFAS in the Mississippi River extend far beyond ecological harm. Downstream communities that depend on the river for drinking water face heightened risks of long-term exposure to these chemicals. Studies have linked PFAS exposure to health issues such as cancer, developmental delays in children, and immune system suppression (Grandjean & Clapp, 2015). Efforts to address PFAS contamination in the Mississippi River include enhanced monitoring programs and targeted cleanup initiatives, but the sheer size and complexity of the river system make comprehensive remediation a daunting task.

Great Lakes

The Great Lakes, which collectively hold 20% of the world's freshwater, are increasingly recognized as hotspots for PFAS contamination. Industrial cities surrounding the lakes, such as Detroit, Cleveland, and Chicago, have historically discharged PFAS-containing waste into these water bodies. This legacy pollution, combined with ongoing industrial discharges, has contributed to elevated PFAS levels in the lakes. Additionally, atmospheric deposition plays a significant role, as volatile PFAS compounds are transported through air currents and eventually settle into the lakes via precipitation (Cousins et al., 2019).

Urban runoff is another major contributor to PFAS contamination in the Great Lakes. Rainwater washing over city streets, parking lots, and industrial zones collects PFAS residues from consumer products and construction materials, funneling these chemicals into stormwater systems and, ultimately, into the lakes. These pollutants have been detected not only in the water but also in sediments, where they accumulate and act as long-term reservoirs of contamination.

One of the most concerning aspects of PFAS in the Great Lakes is its impact on aquatic life and human health. Studies have documented high levels of PFAS in fish species, leading to bioaccumulation throughout the food chain. This has raised alarms about human exposure through seafood consumption, particularly among communities that rely on fishing for subsistence or cultural practices. Public health advisories have been issued to limit fish consumption in certain areas, but addressing the broader issue of PFAS contamination requires more robust interventions. Efforts to mitigate PFAS pollution in the Great Lakes include enhanced monitoring programs, public education campaigns, and the exploration of advanced water treatment technologies. However, the sheer scale of the contamination and the interconnected nature of the lakes present significant challenges to achieving lasting solutions.

Colorado River

The Colorado River, a vital water source for over 40 million people in the western United States, is also grappling with PFAS contamination. This iconic river serves as a lifeline for agricultural irrigation, industrial processes, and urban water supplies, making its contamination a pressing concern. PFAS pollution in the Colorado River has been linked to multiple sources, including agricultural runoff, industrial discharges, and contamination from nearby military bases. For example, the use of aqueous film-forming foam (AFFF) at military installations along the river basin has contributed to elevated PFAS levels in surrounding groundwater and surface water (Nguyen et al., 2020).

The agricultural sector plays a significant role in the contamination of the Colorado River. PFAS are often introduced into agricultural systems through the application of biosolids and irrigation with contaminated water. These chemicals then leach into the river during runoff events, perpetuating their presence in the waterway. Industrial activities near the river further exacerbate the problem, with facilities discharging PFAS-containing waste into tributaries that feed into the Colorado River.

The stakes are particularly high for cities like Las Vegas, Phoenix, and Los Angeles, which rely heavily on the Colorado River for drinking water. Advanced treatment technologies, such as reverse osmosis and granular activated carbon filtration, have been implemented in some areas to reduce PFAS concentrations in treated water. However, these solutions are both resource- and energy-intensive, limiting their widespread adoption. Additionally, the cost of implementing such technologies often falls on municipal water utilities and, by extension, consumers, creating economic burdens for affected communities (Grandjean & Clapp, 2015).

Efforts to address PFAS contamination in the Colorado River include stricter regulations on industrial discharges, enhanced agricultural practices to minimize runoff, and increased federal funding for water treatment and remediation. However, the interconnected nature of the river basin and its extensive use across multiple states complicate efforts to achieve a comprehensive solution. The Colorado River's challenges reflect the broader difficulties of managing PFAS in shared water resources and highlight the need for coordinated regional and federal responses to tackle this persistent threat (Sims et al., 2025).

Role of Military Bases, Manufacturing Plants, and Landfills

Military bases across the United States are among the most significant contributors to PFAS contamination, primarily due to the widespread use of aqueous film-forming foam (AFFF) in firefighting training exercises and emergency response scenarios. AFFFs contain high concentrations of PFAS due to their effectiveness in suppressing petroleum-based fires. Over decades, these foams have been used extensively on military installations, resulting in significant PFAS contamination in groundwater and nearby surface water bodies. Notable examples include Naval Air Station Pensacola in Florida and Peterson Air Force Base in Colorado, where groundwater samples have revealed PFAS concentrations far exceeding health advisory limits set by the U.S. Environmental Protection Agency (Grandjean & Clapp, 2015).

The Department of Defense (DoD) has acknowledged the severity of this issue, with a 2021 report estimating that at least 700 military sites are contaminated with PFAS (U.S. DoD, 2021). The agency has allocated billions of dollars for cleanup efforts and has initiated groundwater monitoring and treatment programs. However, remediation efforts are slow and costly, often involving advanced techniques like granular activated carbon (GAC) filtration or reverse osmosis systems. The complexity and scale of contamination at military sites highlight the challenges of addressing legacy PFAS pollution effectively.

Additionally, the contamination often extends beyond military property, impacting nearby civilian communities that rely on the same groundwater and surface water sources. For instance, communities near Wright-Patterson Air Force Base in Ohio have reported PFAS contamination in their drinking water, leading to lawsuits and calls for stricter regulations (Nguyen et al., 2020). This underscores the broader societal implications of PFAS contamination originating from military bases.

Manufacturing plants that produce or use PFAS in their processes are another major source of environmental contamination. Companies like DuPont and 3M have long been associated with PFAS pollution due to their production of these chemicals for industrial and consumer products. For decades, these companies discharged untreated PFAS waste directly into nearby rivers, streams, and soil. High-profile cases, such as the contamination of the Ohio River near Parkersburg, West Virginia, by DuPont's Washington Works facility, have brought attention to the environmental and health risks associated with these practices (Bilott, 2019).

Legal actions and regulatory scrutiny have forced these manufacturers to address their role in PFAS contamination. For example, 3M agreed to a $850 million settlement with the state of Minnesota in 2018 to fund water quality projects addressing PFAS pollution. Similarly, DuPont has faced multiple lawsuits and class-action settlements for contamination in communities surrounding their manufacturing plants. Despite these efforts, legacy pollution from decades of unchecked discharges continues to pose risks to nearby ecosystems and human populations (Grandjean & Clapp, 2015).

Moreover, small- and medium-sized manufacturing facilities, which often lack the resources to implement advanced wastewater treatment technologies, contribute to the ongoing release of PFAS into the environment. This widespread industrial use of PFAS highlights the need for stricter regulations and technological innovations to reduce contamination at its source.

Landfills play a critical role in PFAS contamination due to the disposal of consumer products containing these chemicals. Items such as food packaging, textiles, carpeting, and electronics frequently end up in landfills at the end of their lifecycle. Over time, PFAS leach out of these products into landfill leachate—a liquid byproduct of waste decomposition. This leachate often infiltrates surrounding soil and groundwater or is transported to municipal wastewater treatment plants, which are typically ill-equipped to remove PFAS effectively (Lang et al., 2017).

Studies have shown that municipal landfill leachate can contribute significant PFAS loads to nearby surface waters. For example, Lang et al. (2017) estimated that PFAS concentrations in landfill leachate range from 1,000 to 10,000 parts per trillion (ppt), far exceeding safety thresholds for drinking water. These findings highlight the urgent need for improved waste management practices, including the treatment of leachate before discharge and the development of PFAS-free alternatives for consumer products.

In addition to municipal landfills, construction and demolition debris landfills are emerging as significant sources of PFAS contamination. Building materials, such as treated wood, roofing membranes, and flooring materials, often contain PFAS, which can leach into the environment when disposed of improperly (Domingo et al., 2021). Addressing this issue requires a comprehensive approach that includes better waste segregation, stricter landfill design standards, and advanced technologies for leachate treatment.

Ultimately, PFAS contamination in the United States is not just an environmental challenge, it is a crisis of national urgency. The widespread infiltration of these "forever chemicals" into the nation's rivers, lakes, and groundwater exemplifies a failure to safeguard our most vital resource: clean water. From the Mississippi River to the Great Lakes and the Colorado River, PFAS pollution threatens ecosystems, disrupts livelihoods, and endangers public health. The pervasive contributions from military bases, manufacturing plants, and landfills underscore the interconnected nature of this issue and the scale of its impact.

Solving this crisis demands bold and immediate action. Stricter regulations must not be optional but mandatory, holding industries and institutions accountable for decades of environmental negligence. Cutting-edge treatment technologies must be prioritized, no matter the cost, to restore contaminated waters and protect future generations. Beyond technological solutions, this is a moral imperative, a responsibility to address the cumulative harm inflicted on communities and ecosystems.

The urgency of this moment cannot be overstated. PFAS contamination is a battle we cannot afford to lose. With coordinated efforts between government agencies, industries, scientists, and communities, we have the tools to turn the tide on this crisis. It is time to rewrite the legacy of PFAS, not as a story of unchecked pollution, but as a defining moment where the United States rose to protect its water, its people, and its future.

Chapter 7

PFAS in Groundwater vs. Surface Water

It is obvious that PFAS contamination is pervasive in both surface water and groundwater systems, but the dynamics of contamination and the associated challenges vary significantly between these two types of water sources. While surface waters such as rivers, lakes, and streams are more visible and subject to direct pollution, groundwater systems often harbor contamination that is hidden but equally harmful. Understanding the differences between PFAS behavior in surface and groundwater, as well as their interactions, is essential for developing effective remediation strategies. This chapter explores the contrasting dynamics of PFAS in these water systems and examines the complexities of addressing contamination in aquifers.

Contamination Dynamics Between Surface and Ground Water

Surface water contamination with PFAS is a prominent and visible environmental issue, as these chemicals are often introduced directly into rivers, lakes, and streams through various human activities. Industrial discharges represent a significant source, as factories producing or using PFAS release untreated wastewater into nearby surface waters. Stormwater runoff exacerbates the problem by washing PFAS from paved surfaces, such as roads and parking lots, into storm drains and waterways, further spreading contamination. Agricultural activities contribute as well, with PFAS entering surface waters through runoff from fields treated with biosolids or irrigation water containing these chemicals. Municipal wastewater treatment plants, which are not designed to remove PFAS effectively, also discharge effluents containing these persistent chemicals into surface water systems. The unique chemistry of PFAS, particularly their strong carbon-fluorine bonds, allows them to persist in these environments, resisting natural degradation processes. As surface waters often act as conduits, PFAS contamination is carried downstream, affecting ecosystems and communities far removed from the original sources of pollution. For instance, the Mississippi River has become a significant repository for PFAS due to a combination of industrial, agricultural, and municipal inputs along its banks, creating widespread impacts on both aquatic ecosystems and human populations reliant on the river for drinking water (Schultz et al., 2006).

In contrast, groundwater contamination by PFAS occurs less visibly but poses equally severe risks. PFAS enter groundwater systems through pathways such as leaching from landfills, the application of PFAS-laden biosolids as fertilizers, and spills from industrial sites or firefighting foam usage. The subsurface movement of PFAS is influenced by the physical and chemical properties of the soil and aquifer materials. For example, sandy soils, which are more porous, allow PFAS to migrate more rapidly and over greater distances. Conversely, clay-rich soils or organic matter can slow PFAS movement by adsorbing these chemicals, although this may only temporarily limit their spread. Regardless of the substrate, the chemical stability of PFAS ensures that once they infiltrate aquifers, they can persist for decades, creating long-term contamination risks for drinking water supplies. This persistence is particularly concerning for rural and urban areas reliant on groundwater as their primary source of potable water (Nguyen et al., 2020).

One of the most critical differences between surface water and groundwater contamination is the visibility and frequency of monitoring. Surface waters, being more accessible, are regularly tested for pollutants, often allowing for earlier detection and intervention. In contrast, groundwater contamination often remains hidden, as PFAS can infiltrate aquifers undetected and persist for years or even decades before being discovered. This hidden nature of groundwater pollution makes it particularly insidious, as communities may unknowingly consume contaminated water over extended periods. Groundwater contamination is often identified only after extensive testing of wells or municipal water supplies, typically in response to observed health impacts or environmental studies. By the time PFAS contamination is confirmed, significant exposure may already have occurred, necessitating costly remediation and long-term monitoring efforts. This disparity in visibility underscores the need for proactive groundwater monitoring and stricter controls to prevent PFAS from infiltrating these critical resources.

Interactions Between Groundwater and Surface Water Systems

Groundwater and surface water systems are intricately interconnected, creating pathways for PFAS contamination to move seamlessly between the two. When groundwater discharges into rivers, lakes, and streams, it can introduce PFAS that have infiltrated the subsurface through industrial activities, landfill leachate, or agricultural runoff. These contaminants, once in surface waters, can spread further downstream, affecting ecosystems and communities that rely on these water bodies for drinking water, recreation, and irrigation. Conversely, surface water infiltration into aquifers can transfer PFAS into groundwater systems. This process is particularly common in regions with high rainfall, porous soils, or extensive irrigation practices, where surface water easily percolates into the subsurface, carrying contaminants along with it. Such bidirectional interactions create a cyclical contamination pattern that complicates efforts to isolate and remediate PFAS pollution effectively.

A striking example of this interplay is the Cape Fear River Basin in North Carolina. In this region, PFAS from industrial discharges, particularly from facilities producing fluorinated compounds, infiltrated the surrounding groundwater. Over time, this contaminated groundwater discharged back into the river system, perpetuating a cycle of pollution. This interaction underscores the challenges of addressing PFAS contamination, as tackling surface water pollution without considering groundwater sources, or vice versa, can result in incomplete remediation and continued environmental and human health risks (Cousins et al., 2019). The Cape Fear River case has become a focal point for researchers and policymakers, highlighting the need for comprehensive monitoring programs that address both groundwater and surface water contamination simultaneously.

Hydrological conditions play a critical role in determining the nature and extent of interactions between groundwater and surface water systems. In regions with high water tables, groundwater is more likely to discharge into nearby rivers and lakes, creating direct pathways for PFAS to migrate from aquifers into surface water. This is particularly common in wet or flood-prone areas, where the close proximity of groundwater to surface water bodies facilitates contamination transfer. On the other hand, in arid or semi-arid regions, where surface water sources are scarce, the reliance on groundwater for drinking water, agricultural irrigation, and industrial processes amplifies the risks associated with PFAS contamination in aquifers. In such areas, surface water infiltration—whether through artificial recharge practices or natural seepage—can introduce PFAS into groundwater supplies, compounding the contamination challenge.

The interconnected nature of groundwater and surface water also means that remediation strategies must account for both systems. For instance, cleaning up PFAS in a river without addressing the contaminated groundwater feeding into it is unlikely to yield long-term success. Similarly, removing PFAS from an aquifer while ignoring the contributions of infiltrating surface water can result in recontamination. Effective remediation therefore requires an integrated approach that considers the dynamic interactions between these two systems. Advanced monitoring technologies, such as high-resolution mass spectrometry and hydrological modeling, can help identify these interactions and guide more effective interventions. By adopting a holistic perspective, policymakers and environmental managers can better address the complexities of PFAS contamination and develop strategies that protect both surface water and groundwater resources.

Challenges in Remediating Contaminated Aquifers

Remediating PFAS-contaminated aquifers presents a range of unique and daunting challenges, owing to the inherent complexities of groundwater systems and the persistent chemical properties of PFAS. Unlike many other contaminants, PFAS are resistant to natural degradation processes due to their strong carbon-fluorine bonds, which make them chemically and thermally stable. Traditional water treatment methods, such as standard filtration and biological processes, have proven largely ineffective against PFAS, as these techniques are not designed to target the unique molecular structure of these "forever chemicals." As a result, more advanced treatment methods, such as granular activated carbon (GAC) and reverse osmosis, are required. While these technologies are more effective at capturing or filtering PFAS, they come with significant downsides, including high costs, energy requirements, and the challenge of managing the concentrated PFAS-laden waste byproducts they generate (Grandjean & Clapp, 2015).

One of the most significant challenges in addressing PFAS contamination in aquifers is the scale and depth of the problem. Groundwater contamination often spans vast geographic areas and penetrates deep into subsurface layers, making cleanup efforts logistically complex and resource intensive. Unlike surface water contamination, which can often be addressed at the source or downstream, aquifer remediation typically requires a labor-intensive approach known as "pump-and-treat." This method involves extracting contaminated groundwater, treating it above ground using technologies like GAC or reverse osmosis, and reinjecting the treated water back into the aquifer. While effective in some cases, pump-and-treat systems are time-consuming and expensive to operate. Furthermore, this approach is often impractical for addressing contamination in large or deep aquifers, where the sheer volume of water and the difficulty of accessing deeper layers can render cleanup efforts unfeasible (Nguyen et al., 2020).

Emerging technologies, such as in situ chemical oxidation and electrochemical treatment, are being explored as potential solutions to the challenges of PFAS remediation. In situ methods aim to treat PFAS contamination directly within the aquifer, reducing the need for water extraction and external treatment. For example, in situ chemical oxidation involves injecting reactive chemicals into the aquifer to break down PFAS molecules into less harmful byproducts. Electrochemical treatment, on the other hand, uses electrical currents to degrade PFAS compounds in the subsurface. While these approaches hold promise, they remain in the experimental stages and face significant technical, regulatory, and financial barriers to widespread adoption. Moreover, there are concerns about the potential formation of harmful byproducts during the degradation process, as well as the long-term effectiveness of these methods in complex hydrogeological settings (Liu et al., 2021).

The long-term nature of PFAS contamination further complicates remediation efforts. Even if new sources of contamination are eliminated, the legacy pollution in aquifers can continue to impact water quality for decades or even centuries. PFAS tend to bind to soil particles and organic matter in the subsurface, creating reservoirs of contamination that can slowly leach into groundwater over time. This persistence underscores the importance of implementing proactive measures to address PFAS contamination at its source. Preventing

PFAS releases through stricter industrial regulations, improved waste management practices, and the development of PFAS-free alternatives is critical to reducing future contamination. Additionally, more robust monitoring systems are needed to detect and track PFAS in groundwater early, enabling faster responses to emerging contamination threats.

Remediating PFAS-contaminated aquifers is a monumental task that requires a multifaceted approach. Advances in treatment technologies must be coupled with preventative measures and policy interventions to address both current and future challenges. As research continues to improve our understanding of PFAS behavior and remediation strategies, a coordinated effort among governments, industries, and scientists will be essential to safeguarding groundwater resources and protecting public health.

Finally, PFAS contamination in surface water and groundwater systems represents one of the most complex and urgent environmental challenges of our time. The stark differences between the visible, often immediate impacts on surface waters and the hidden, insidious threats posed by groundwater underscore the multifaceted nature of this crisis. Yet, the interconnectedness of these systems highlights an even greater challenge: contamination in one inevitably affects the other, creating a vicious cycle that demands holistic and innovative solutions. Addressing PFAS is not just a matter of advancing technology; it requires a global commitment to research, regulation, and proactive prevention. This is not merely a scientific or policy issue, it is a moral imperative. The health of our water resources underpins the well-being of ecosystems, communities, and future generations. Ensuring safe, clean water free of these "forever chemicals" is a responsibility we must collectively embrace, turning knowledge into action to secure a sustainable and equitable future for all.

Chapter 8

Health Risks of PFAS

Forever chemicals represent one of the most pervasive and environmentally ubiquitous threats to human health and the environment in the modern era. With their unique chemical stability and resistance to degradation, these substances have infiltrated nearly every aspect of daily life, our water, food, air, and even our bodies. Yet, the health impacts of PFAS exposure are only beginning to be fully understood. From their links to cancer and immune suppression to the reproductive and developmental challenges they pose, PFAS contamination has become a silent crisis affecting communities worldwide. This chapter delves into the health risks associated with PFAS exposure, exploring the scientific evidence, community-level impacts, and the urgent need for action to address this growing public health emergency.

Introduction to PFAS and Health

PFAS, or per- and polyfluoroalkyl substances, are a class of over 12,000 synthetic chemicals that have become ubiquitous in modern life due to their unique properties, such as water and oil repellency, thermal stability, and resistance to chemical degradation. These properties are derived from the presence of strong carbon-fluorine

99

bonds, which are among the strongest bonds in chemistry, making PFAS nearly indestructible under natural conditions. This remarkable durability, while advantageous in industrial and consumer applications, has earned PFAS the nickname "forever chemicals," as they persist indefinitely in the environment and accumulate in living organisms over time (Grandjean & Clapp, 2015).

PFAS are particularly concerning for human health because of their bioaccumulative and biomagnifying characteristics. Unlike many pollutants that are metabolized and excreted, PFAS can remain in the human body for years, building up in organs such as the liver, kidneys, and blood with repeated exposure. These chemicals enter the body through various pathways, including contaminated drinking water, food grown in PFAS-impacted soil, air near industrial facilities, and everyday items like nonstick cookware, waterproof clothing, and food packaging (Nguyen et al., 2020).

Research has shown that even low levels of PFAS exposure can lead to serious health consequences. Epidemiological studies have linked PFAS to a range of health issues, including cancer, immune system suppression, thyroid dysfunction, and reproductive and developmental problems. For example, studies conducted in heavily contaminated areas, such as Parkersburg, West Virginia, where DuPont's discharges of perfluorooctanoic acid (PFOA) polluted water supplies, revealed elevated risks of kidney and testicular cancers among residents (Bilott, 2019).

Adding to the urgency is the fact that PFAS contamination is not confined to industrial areas or highly polluted regions; it has become a global issue, with traces found in remote ecosystems, such as the Arctic, and even in the blood of newborns, highlighting its far-reaching and insidious nature (Domingo et al., 2021). The omnipresence and resilience of PFAS make them a critical public health concern that demands immediate and coordinated action from policymakers, industry leaders, researchers, and affected communities alike. Failing to address this issue could lead to lasting consequences for human health and environmental sustainability.

The persistence and bioaccumulative nature of PFAS underscore the need for enhanced monitoring, stricter regulatory standards, and innovative solutions to mitigate exposure and protect vulnerable populations. The increasing body of scientific evidence pointing to their harmful effects has already spurred advocacy efforts and regulatory action in some regions, but much work remains to address the global scale of this crisis.

Links Between PFAS Exposure and Health Outcomes

Exposure to PFAS has been linked to a wide range of health outcomes, including cancer, reproductive and developmental issues, endocrine disruption, immune system suppression, and cardiovascular impacts. These health risks stem from the bioaccumulative and persistent nature of PFAS, which allows them to remain in the body and disrupt vital biological processes over time.

Cancer Risks

PFAS exposure has been strongly linked to an increased risk of specific cancers, particularly kidney and testicular cancers. Epidemiological studies, such as those conducted on residents of Parkersburg, West Virginia, revealed a clear association between long-term exposure to perfluorooctanoic acid (PFOA), a type of PFAS, and higher incidences of these cancers (Bilott, 2019). Residents in the area experienced chronic exposure through contaminated drinking water, leading to widespread health issues. Industrial workers who directly handled PFAS chemicals have similarly exhibited increased cancer rates, further supporting the carcinogenic potential of these substances.

Research has shown that PFAS may promote cancer by interfering with cellular processes, including DNA repair mechanisms and the regulation of cell growth. For instance, studies suggest that PFAS may activate pathways involved in tumor growth, such as the peroxisome proliferator-activated receptor (PPAR) pathway, which has been implicated in kidney and testicular cancers (Nguyen et al., 2020). Additionally, the persistent and bioaccumulative nature of PFAS allows these chemicals to remain in the body for years, increasing the risk of long-term cellular damage and the subsequent development of malignancies.

Reproductive and Developmental Issues

PFAS disrupt reproductive systems by interfering with hormone regulation, significantly impacting fertility, pregnancy outcomes, and fetal development. Exposure to PFAS during pregnancy has been linked to adverse outcomes such as low birth weight, preterm births, and developmental delays in children. These effects are particularly concerning because PFAS can cross the placental barrier, exposing fetuses to the chemicals during critical stages of development (Grandjean & Clapp, 2015).

Studies have also shown that PFAS exposure can impair fertility in both men and women. In women, PFAS are associated with irregular menstrual cycles, decreased ovarian reserve, and disruptions in hormone levels essential for reproduction, such as estrogen and progesterone. In men, PFAS exposure has been linked to reduced sperm quality, lower testosterone levels, and impaired testicular function, all of which can hinder fertility (Domingo et al., 2021).

Furthermore, PFAS exposure may affect pubertal development in adolescents, with studies suggesting earlier onset of puberty in girls and delayed reproductive maturation in boys. These findings highlight the far-reaching implications of PFAS on reproductive health and underscore the need for preventive measures, especially for vulnerable populations such as pregnant women and children.

Endocrine Disruption

The endocrine-disrupting properties of PFAS have raised significant concerns due to their ability to interfere with hormonal systems that are critical for growth, metabolism, and overall health. These chemicals are known to disrupt thyroid hormone production, which is essential for regulating metabolism and supporting brain development, particularly in infants and young children. Altered thyroid hormone levels caused by PFAS exposure can lead to conditions such as hypothyroidism, which can have long-term effects on cognitive development, energy levels, and physical growth in young individuals (Domingo et al., 2021).

For vulnerable populations such as pregnant women and young children, the effects of PFAS-induced thyroid disruption are particularly alarming. During pregnancy, thyroid hormones play a crucial role in fetal brain development, and any disruption can result in developmental delays, lower IQ levels, or even congenital hypothyroidism. For young children, PFAS exposure could lead to slowed growth, delayed puberty, and challenges in reaching developmental milestones.

In addition to thyroid disruption, PFAS exposure has been linked to broader metabolic issues, including obesity and type 2 diabetes. Research indicates that PFAS may alter glucose metabolism and impair insulin regulation, increasing the risk of diabetes in both adults and children. Children exposed to higher PFAS levels may experience greater susceptibility to weight gain and metabolic disorders, potentially setting them up for lifelong health challenges, such as cardiovascular disease and chronic inflammation (Grandjean & Clapp, 2015).

The impact of PFAS on the endocrine system extends beyond metabolic health. Disruption of adrenal gland function, which impairs the production of cortisol and other essential hormones, can increase stress levels and weaken the body's ability to respond to physical and emotional stressors. For immuno-compromised individuals, such disruptions can exacerbate pre-existing conditions, leaving them more vulnerable to infections and chronic illnesses. The cumulative effects of these disruptions underscore the wide-ranging health risks posed by PFAS and their potential to exacerbate chronic conditions over time.

Immune System Effects

PFAS exposure has been shown to suppress the immune system, posing severe risks for young children, the elderly, and immunocompromised individuals. This immunosuppressive effect has been particularly evident in studies on children, where higher PFAS levels have been linked to lower antibody responses to routine vaccinations, such as tetanus and diphtheria. These findings suggest that PFAS can weaken vaccine efficacy, leaving children more vulnerable to preventable diseases and limiting the effectiveness of public health immunization programs (Grandjean & Clapp, 2015).

Young children, whose immune systems are still developing, are especially at risk of long-term health consequences from immune suppression caused by PFAS exposure. A weakened immune system not only increases their susceptibility to infections but also prolongs recovery times, leading to more severe outcomes from common illnesses. For immunocompromised individuals, such as those undergoing treatment for autoimmune disorders or cancer, PFAS exposure could further compromise immune function, increasing the likelihood of complications and hospitalizations (Nguyen et al., 2020).

Moreover, emerging studies suggest a potential link between PFAS exposure and autoimmune diseases, such as lupus, multiple sclerosis, and rheumatoid arthritis. These conditions, where the immune system mistakenly attacks healthy tissues, may be triggered or exacerbated by PFAS-induced immune dysregulation. Children and individuals with genetic predispositions to autoimmune disorders may face heightened risks when exposed to elevated PFAS levels.

The immunotoxin effects of PFAS extend beyond individual health, affecting entire communities exposed to high levels of contamination. For example, communities with PFAS-contaminated water supplies often experience higher rates of illness, increased healthcare demands, and greater financial burdens. These findings emphasize the critical need for measures to reduce PFAS exposure and protect vulnerable populations, particularly children, the elderly, and those with pre-existing health conditions. Proactive efforts in monitoring, reducing exposure, and developing public health strategies are essential to mitigate the far-reaching impacts of PFAS on immune health.

Cardiovascular and Liver Impacts

PFAS exposure has been increasingly linked to adverse cardiovascular and liver health outcomes, with significant implications for long-term public health. One of the most well-documented effects of PFAS is their association with elevated cholesterol levels, a major risk factor for cardiovascular disease. Epidemiological studies have shown that individuals exposed to higher levels of PFAS, particularly PFOA and PFOS, often exhibit increased total cholesterol and low-density lipoprotein (LDL) cholesterol levels (Grandjean & Clapp, 2015). This elevated cholesterol can contribute to the development of atherosclerosis, a condition where plaque builds up in the arteries, leading to an increased risk of heart attack, stroke, and other cardiovascular complications.

Beyond cholesterol, PFAS have been linked to hypertension and disruptions in blood pressure regulation. Some studies suggest that PFAS exposure may impair the function of endothelial cells, which line the blood vessels, leading to reduced vascular elasticity and increased risk of hypertension (Domingo et al., 2021). These cardiovascular effects are particularly concerning for populations already at risk, such as individuals with diabetes, obesity, or genetic predispositions to heart disease.

PFAS also pose significant risks to liver health. Research has identified a strong connection between PFAS exposure and liver toxicity, including the development of non-alcoholic fatty liver disease (NAFLD). NAFLD is characterized by the abnormal accumulation of fat in the liver, which can progress to more severe conditions such as non-alcoholic steatohepatitis (NASH), liver fibrosis, and eventually cirrhosis. Studies have found that PFAS disrupt the liver's ability to regulate lipid metabolism, leading to increased fat accumulation and inflammation (Nguyen et al., 2020).

The impacts of PFAS on the liver extend beyond NAFLD. PFAS exposure has been associated with elevated liver enzymes, such as alanine aminotransferase (ALT) and aspartate aminotransferase (AST), which are markers of liver damage. These changes indicate stress and injury to liver cells, even at low levels of PFAS exposure. The liver, as a critical organ for detoxification and metabolism, becomes particularly vulnerable to the bioaccumulative nature of PFAS, which can remain in the body for years and perpetuate long-term damage.

Children and individuals with pre-existing liver conditions are especially vulnerable to the hepatotoxic effects of PFAS. For children, disruptions to liver function during critical stages of growth and development can have lifelong consequences, including impaired metabolic regulation and increased susceptibility to obesity and diabetes. For those with existing liver conditions, PFAS exposure may accelerate disease progression and worsen overall outcomes.

These findings underscore the urgent need for preventive measures to address PFAS contamination and mitigate their cardiovascular and liver impacts. Policies aimed at reducing PFAS exposure in drinking water, food, and consumer products, combined with advances in medical screening and treatment, are essential for protecting public health from the insidious effects of these "forever chemicals."

Community-Level Impacts

The impacts of PFAS contamination extend beyond individual health outcomes, affecting entire communities on a socioeconomic and environmental scale. From rural areas with limited access to clean water to industrialized regions burdened by decades of pollution, PFAS contamination disproportionately affects vulnerable populations, exacerbating existing disparities. Communities near military bases, manufacturing plants, and landfills often face higher exposure risks, creating an urgent need for environmental justice and equitable solutions. Addressing the community-level impacts of PFAS requires not only technological advancements but also systemic changes to prioritize the health and well-being of affected populations.

Disparities in Exposure

Environmental justice concerns are a central issue in PFAS contamination, as low-income and minority communities often face a disproportionate burden of exposure. These groups are more likely to live near industrial facilities, military bases, or landfills, which are common sources of PFAS pollution. Limited access to clean water infrastructure and healthcare exacerbates their vulnerability, leaving residents with fewer resources to address contamination and its health impacts. Children and pregnant women in these communities face heightened risks, as exposure during critical developmental periods can result in long-term health consequences, including low birth weight, developmental delays, and immune system suppression (Grandjean & Clapp, 2015).

Additionally, the intersection of poverty and pollution creates a cycle of vulnerability. Low-income families may rely on private wells, which are often unregulated and more susceptible to PFAS contamination. These communities also face greater challenges in advocating for regulatory action or accessing legal remedies. This disparity underscores the need for targeted interventions to protect vulnerable populations from the health and socioeconomic impacts of PFAS exposure.

Case Studies of Affected Communities

Across the United States, communities have experienced firsthand the devastating impacts of PFAS contamination. From industrial pollution to military site discharges, these cases illustrate the widespread nature of the crisis and its profound effects on public health and the environment. By examining specific examples, such as Parkersburg, West Virginia, the Cape Fear River Basin, and Oscoda, Michigan, we gain a deeper understanding of the challenges faced by affected populations and the urgent need for comprehensive solutions.

Parkersburg, West Virginia

Parkersburg, West Virginia, serves as one of the most infamous examples of PFAS contamination in the United States. For decades, the DuPont plant discharged PFOA (a type of PFAS) into local water supplies, contaminating both the environment and the community's health. Residents experienced elevated rates of kidney and testicular cancer, thyroid disorders, and immune system deficiencies. The case gained national attention through legal battles, including a class-action lawsuit that revealed DuPont's negligence in addressing the contamination (Bilott, 2019). The Parkersburg incident highlights the devastating health consequences of unchecked industrial pollution and the challenges communities face in holding corporations accountable.

Cape Fear River Basin, North Carolina

In North Carolina's Cape Fear River Basin, chronic PFAS contamination from industrial discharges has created a public health crisis. Chemours, a spin-off company from DuPont, released GenX, a replacement for PFOA, into the river for years, contaminating drinking water for over 300,000 residents. Studies have linked exposure to developmental delays, hormonal imbalances, and weakened immune responses, particularly among children and pregnant women. Despite growing public pressure and regulatory action, the legacy of contamination continues to impact on the health and well-being of the region's residents. This case underscores the need for stricter regulations and advanced water treatment solutions to protect vulnerable communities.

Oscoda, Michigan

Another notable case is Oscoda, Michigan, where the former Wurtsmith Air Force Base contributed to severe PFAS contamination in groundwater through the use of firefighting foams containing PFAS. Local residents relying on private wells were exposed to dangerous levels of PFAS, leading to health concerns such as liver damage, high cholesterol, and immune dysfunction. Oscoda's experience highlights the ongoing challenges of cleaning up PFAS contamination near military sites, where funding and remediation efforts often fall short of community needs.

Socioeconomic and Mental Health Impacts

The economic burden of PFAS contamination is immense, with communities bearing the brunt of rising healthcare costs, property devaluation, and diminished productivity due to illness. Families affected by PFAS exposure often face significant out-of-pocket expenses for medical treatments, water filtration systems, and legal battles, compounding the financial strain on already vulnerable populations. For municipalities, the cost of upgrading water treatment facilities to remove PFAS often runs into millions of dollars, diverting resources from other essential services.

The mental health toll on residents in contaminated areas is equally significant. Chronic stress and anxiety are prevalent as individuals worry about the long-term health risks posed by PFAS exposure, particularly for their children and future generations. Feelings of helplessness and frustration are common, as many communities face lengthy delays in remediation efforts and limited accountability from polluting entities. The stigma associated with living in a contaminated area can also lead to social isolation and diminished community morale (Domingo et al., 2021).

Moreover, the devaluation of property in PFAS-contaminated areas poses another layer of socioeconomic impact. Homeowners often find it difficult to sell their properties, trapping families in affected areas and further exacerbating financial instability. This combination of health, economic, and psychological burdens underscores the far-reaching consequences of PFAS contamination and the critical need for comprehensive policies that address both environmental cleanup and community support.

PFAS Contamination in Drinking Water

PFAS contamination in drinking water represents one of the most pressing public health challenges associated with these "forever chemicals." With drinking water serving as a primary exposure route, millions of people across the globe are at risk of adverse health effects. Examining the pathways of contamination, affected areas, and the long-term consequences highlights the critical need for improved water management and regulatory action to protect public health.

PFAS frequently infiltrate municipal water supplies through a variety of pathways, including industrial discharges, runoff from firefighting foam applications, and leachate from landfills. These chemicals can also enter water systems through the discharge of treated wastewater, as conventional water treatment facilities are not designed to effectively remove PFAS. Municipal systems are particularly at risk when located near industrial sites or military bases where PFAS use has been prevalent for decades.

Private wells in rural areas face an even greater vulnerability. Unlike municipal systems, private wells often lack routine testing and treatment, leaving residents unknowingly exposed to PFAS. For example, rural communities near landfills or agricultural areas where biosolids are applied as fertilizer are at significant risk of contamination. The lack of regulation and resources for monitoring private wells exacerbates the exposure risks for these populations, disproportionately impacting low-income and marginalized groups (Schultz et al., 2006).

Long-Term Health Consequences

Chronic exposure to PFAS-contaminated drinking water has been linked to a wide range of long-term health effects. Epidemiological studies have identified associations between PFAS exposure and cancers, particularly kidney and testicular cancers, as well as non-cancerous conditions like immune suppression, thyroid disorders, and reproductive health issues. These outcomes are often the result of prolonged exposure to low concentrations of PFAS, a hallmark of drinking water contamination.

One of the most troubling aspects of PFAS-related health effects is the latency period for diseases like cancer, which can take years or even decades to manifest. This delayed onset makes it difficult to establish direct links between exposure and health outcomes, complicating efforts to address the problem through public health initiatives. For example, communities that have consumed PFAS-contaminated water for decades may only now be experiencing the full extent of the health impacts, overwhelming healthcare systems and creating significant public health challenges (Nguyen et al., 2020).

PFAS have also been shown to impair the immune system, reducing vaccine efficacy and increasing susceptibility to infections. Children and pregnant women are particularly vulnerable, as exposure during critical developmental periods can lead to lasting health consequences, including developmental delays, low birth weight, and long-term immune dysfunction. The enduring presence of PFAS in drinking water underscores the urgency of addressing this issue to prevent further harm to human health.

Case Studies of Drinking Water Contamination

Cases of PFAS contamination in drinking water highlight the widespread and severe impact of these chemicals on communities. From small towns like Hoosick Falls to larger regions like North Carolina's Cape Fear River Basin, these examples reveal the urgent need for robust monitoring, advanced treatment solutions, and stronger regulations to protect public health and ensure clean, safe water for all.

Hoosick Falls, New York

One of the most prominent examples of PFAS contamination in drinking water occurred in Hoosick Falls, New York, where industrial discharges from a local manufacturing facility contaminated the municipal water supply with PFOA. Residents were exposed to dangerously high levels of PFAS for years before the issue was detected. Subsequent health studies revealed elevated cancer rates and other adverse health effects among the population. This case underscores the need for proactive water quality monitoring and swift regulatory action to address contamination (Bilott, 2019).

Flint, Michigan

Although the Flint, Michigan, water crisis is primarily associated with lead contamination, it highlights broader issues of systemic failures in water management that also apply to PFAS contamination. The crisis exposed vulnerabilities in municipal water systems, such as inadequate testing, aging infrastructure, and delayed governmental response. These same weaknesses have been identified in many cases of PFAS contamination, illustrating the need for comprehensive reform in water governance and treatment infrastructure.

North Carolina's Cape Fear River Basin

Communities reliant on the Cape Fear River for drinking water have faced chronic PFAS contamination due to industrial discharges from the Chemours Fayetteville Works facility. Over 300,000 residents have been affected, with studies linking exposure to health effects such as immune suppression, developmental issues, and hormonal disruptions. Despite public pressure and legal actions, remediation efforts have been slow, leaving residents dependent on costly household filtration systems to ensure water safety. This case highlights the need for stricter industrial regulations and advanced treatment technologies to protect water supplies from PFAS contamination.

Regulatory and Research Gaps in Addressing Health Risks

The health risks associated with PFAS contamination are well-documented, yet significant regulatory and research gaps continue to hinder efforts to protect public health. The absence of standardized global guidelines, the challenges of linking exposure to specific health outcomes, and the need for long-term, comprehensive studies highlight the urgent need for coordinated action and robust scientific inquiry to address the risks posed by these persistent chemicals.

Lack of Standardized Guidelines

One of the most significant obstacles in addressing PFAS-related health risks is the variability in national and international regulatory limits for PFAS in drinking water. While some countries, such as Sweden and Denmark, have adopted stringent guidelines for PFAS levels, others either lack enforceable standards or have set limits that may not adequately protect public health. For instance, in the United States, the Environmental Protection Agency (EPA) has issued health advisories for PFOA and PFOS at 70 parts per trillion (ppt) in drinking water yet states such as New Jersey and California have established stricter limits, reflecting differing interpretations of the science and acceptable risk (U.S. EPA, 2022).

Globally, the lack of harmonized standards complicates mitigation efforts, particularly in developing countries where resources for water testing and enforcement are limited. This disparity leaves vulnerable populations at higher risk of PFAS exposure, particularly in regions heavily reliant on untreated or minimally treated water supplies. Establishing consistent international guidelines for PFAS levels in drinking water is critical to ensuring equitable protection from exposure (Grandjean & Clapp, 2015).

Challenges in Linking Exposure to Outcomes

The long latency periods associated with PFAS-linked diseases, such as cancer and immune dysfunction, pose a significant challenge for researchers and policymakers. It can take years or even decades for adverse health outcomes to manifest following initial exposure, making it difficult to establish clear causal relationships. Furthermore, PFAS exposure rarely occurs in isolation. Individuals are often simultaneously exposed to a mixture of chemicals and environmental stressors, complicating efforts to attribute specific health effects to PFAS (Domingo et al., 2021).

For example, communities exposed to PFAS through contaminated drinking water often face multiple environmental risks, such as lead or pesticide exposure, which can exacerbate or mask the impacts of PFAS. This multifaceted nature of exposure highlights the need for advanced epidemiological methods and comprehensive data collection to isolate the health effects of PFAS and inform targeted mitigation strategies.

Need for Long-Term Studies

Although considerable research has been conducted on PFOA and PFOS, significant gaps remain in understanding the health impacts of lesser-studied PFAS compounds and their combined effects when present as mixtures. Many PFAS compounds, such as GenX chemicals and short-chain alternatives, have been marketed as safer replacements for long-chain PFAS, yet their health risks are not fully understood.

Ongoing cohort studies, such as the C8 Health Project, have been instrumental in identifying health effects linked to PFAS exposure, but these efforts must be expanded to include a broader range of compounds and populations. Additionally, long-term studies are needed to assess the cumulative risks of PFAS exposure over time, particularly for vulnerable populations such as children, pregnant women, and immunocompromised individuals (Nguyen et al., 2020).

International research collaborations can help fill these gaps by pooling resources, standardizing methodologies, and sharing data across borders. Such initiatives would provide a more comprehensive understanding of PFAS health risks and support the development of evidence-based regulations and interventions.

Addressing the regulatory and research gaps in PFAS management is essential to mitigating their long-term health risks. Establishing standardized guidelines, overcoming challenges in linking exposure to health outcomes, and investing in long-term studies are critical steps in protecting public health. As the global community grapples with the pervasive presence of PFAS, prioritizing research and harmonizing regulatory efforts will be key to reducing exposure and safeguarding future generations.

Preventive Measures and Community Advocacy

The fight against PFAS contamination requires proactive measures to protect public health and empower communities. Improved testing and monitoring, community-driven advocacy, and access to safe water alternatives are essential components of a comprehensive strategy to address the widespread impact of these chemicals. By focusing on prevention and education, stakeholders can work together to mitigate risks and drive meaningful change.

Improved Testing and Monitoring

Expanding PFAS testing in public water systems and private wells is critical for early detection and prevention of health risks. Currently, testing for PFAS is not uniformly required across the United States, leaving many communities unaware of potential contamination in their water supplies. Establishing mandatory testing programs for all water systems, especially in areas near industrial facilities, military bases, or landfills, would provide vital data on PFAS concentrations and help prioritize remediation efforts (Schultz et al., 2006).

In addition to testing water sources, incorporating PFAS screening into health monitoring programs for high-risk populations—such as children, pregnant women, and those living in heavily contaminated areas—can help identify early signs of PFAS-related health issues. This data would enable healthcare providers to implement targeted interventions and improve long-term health outcomes. Advances in analytical technologies, such as high-resolution mass spectrometry, can further enhance the accuracy and scope of PFAS detection, supporting more effective monitoring and regulatory oversight.

Community Education and Advocacy

Community-driven advocacy plays a central role in addressing PFAS contamination. Empowering residents to demand stricter regulations, improved water treatment infrastructure, and enhanced remediation efforts can drive action at local, state, and federal levels. Grassroots organizations and community leaders are often the first to raise awareness of contamination issues and advocate for change, making their involvement essential to long-term solutions (Grandjean & Clapp, 2015).

Public awareness campaigns can also help individuals reduce their personal exposure to PFAS. These initiatives can provide practical advice on avoiding PFAS-containing products, such as nonstick cookware and certain cosmetics, and guide consumers on choosing water filters capable of removing PFAS. Education efforts should prioritize reaching vulnerable populations, including low-income and minority communities, who may face higher exposure risks due to proximity to contamination sources and limited access to safe alternatives.

Access to Safe Water Alternatives

Providing affected communities with access to safe drinking water is a critical step in reducing PFAS exposure. Advanced treatment technologies, such as granular activated carbon (GAC) and reverse osmosis, can effectively remove PFAS from municipal water supplies. However, these solutions are often expensive and may take years to implement on a large scale. In the interim, offering alternative water sources—such as bottled water or point-of-use filters—can provide immediate relief to impacted residents (Nguyen et al., 2020).

Investing in innovative, cost-effective filtration systems for households and small communities can bridge the gap while broader infrastructure improvements are underway. Governments and industry stakeholders must collaborate to fund these initiatives, ensuring equitable access to safe water regardless of socioeconomic status. Programs that subsidize water filtration systems for low-income families or provide temporary water deliveries to contaminated areas are examples of measures that can make a tangible difference.

Ultimately, preventive measures and community advocacy are indispensable in addressing the far-reaching impacts of PFAS contamination. By expanding testing and monitoring efforts, educating and empowering communities, and ensuring access to safe drinking water, stakeholders can reduce exposure risks and protect public health. These proactive steps not only mitigate the immediate effects of PFAS but also lay the foundation for a more sustainable and equitable approach to managing environmental contamination. Addressing PFAS is a collective responsibility that requires collaboration, innovation, and a commitment to prioritizing the well-being of all communities.

PFAS contamination stands as one of the most pressing public health and environmental crises of our time, threatening the well-being of individuals, the integrity of ecosystems, and the trust in the systems that sustain us. The undeniable links between PFAS exposure and severe health outcomes, cancer, reproductive issues, endocrine disruption, and immune suppression, underscore the urgency of addressing this pervasive issue. Disparities in exposure, particularly in low-income and marginalized communities, amplify the ethical imperative for action. This is a challenge that transcends borders, requiring unified efforts from governments, industries, scientists, and communities to enact stringent regulations, advance innovative research, and champion prevention. Access to clean, safe water must be upheld as a universal human right, not a privilege. Tackling PFAS is more than a scientific or regulatory necessity; it is a moral obligation to protect the health of current and future generations and to restore faith in our collective ability to meet the challenges of our shared environment.

Chapter 9

Ecosystem Impacts of PFAS

The ecological consequences of PFAS contamination are profound, reaching every corner of the globe and threatening the health and balance of natural systems. These chemicals, while invisible to the naked eye, leave a significant footprint on aquatic ecosystems, biodiversity, and the very sediment layers that form the foundation of water quality. From the disruption of fisheries to the infiltration of remote and polar regions, PFAS contamination underscores the far-reaching and insidious nature of these "forever chemicals." Understanding these impacts is essential to crafting effective solutions that protect both human and ecological health.

Effects on Aquatic Ecosystems, Biodiversity, and Fisheries

Aquatic ecosystems are among the most vulnerable to PFAS contamination due to the unique properties of these chemicals. While PFAS readily dissolve and persist in water, they interact with organisms across every trophic level. Once introduced into aquatic environments, PFAS bioaccumulate in species such as fish, invertebrates, and amphibians, disrupting their physiological processes, growth, and reproduction. Studies have shown that PFAS

119

exposure leads to decreased fertility, impaired development, and increased mortality in fish populations. This not only affects individual species but can also result in cascading effects on aquatic biodiversity, leading to long-term declines in ecosystem stability and productivity (Domingo et al., 2021; Liu et al., 2021).

The impact of PFAS bioaccumulation extends to larger predators, including birds and marine mammals, which consume contaminated prey. Biomagnification amplifies PFAS concentrations as these chemicals move up the food chain, increasing the risks for apex predators and further destabilizing ecosystems. Research in Arctic ecosystems, for example, has documented high levels of PFAS in polar bears and seals, illustrating how even remote regions are not immune to these effects (Butt et al., 2010; Cousins et al., 2022). This contamination compromises the health and survival of apex species, which play critical roles in maintaining the balance of their ecosystems.

Fisheries face significant challenges due to PFAS contamination. Toxicity in fish populations has led to reduced stocks, which directly impacts the livelihoods of communities that rely on fishing for income and food security. Furthermore, PFAS contamination often renders fish unsafe for human consumption, as chemical levels may exceed safety thresholds established by regulatory agencies. For example, studies in the Great Lakes region have highlighted elevated PFAS concentrations in fish, leading to public health advisories and economic losses for local fisheries (Nguyen et al., 2020; Grandjean et al., 2023).

Additionally, the disruption of aquatic ecosystems due to PFAS contamination has broader implications for ecological health and community well-being. As fish populations decline and biodiversity is compromised, entire food webs are affected, impacting not only aquatic life but also terrestrial species that depend on these ecosystems. The interconnectedness of ecological and human systems underscores the urgent need for targeted interventions to mitigate PFAS pollution, protect aquatic biodiversity, and ensure sustainable fisheries.

PFAS in Sediments and Long-Term Implications for Water Quality

Sediments in rivers, lakes, and estuaries act as long-term reservoirs for PFAS, trapping these chemicals in layers of organic material and mineral particles. PFAS enter sediments through industrial discharge, stormwater runoff, and deposition from the water column, where they can persist for decades due to their chemical stability and resistance to degradation. The strong carbon-fluorine bonds that define PFAS make them particularly resilient, allowing them to remain embedded in sediments for extended periods (Ahrens & Bundschuh, 2014; Liu et al., 2021).

However, sediments are not inert. Environmental disturbances, such as flooding, erosion, dredging, or even changes in pH and temperature, can remobilize PFAS, causing them to re-enter the water column. This cycling between sediments and surface water creates a sustained source of contamination that can continue long after primary pollution sources have been addressed. For example, studies in the Great Lakes have shown that PFAS trapped in sediments can perpetuate contamination, complicating water quality management efforts (Nguyen et al., 2020).

The impacts of PFAS in sediments extend beyond water quality to affect benthic organism species that inhabit sediment layers and play vital roles in nutrient cycling and the aquatic food web. These bottom-dwelling organisms are directly exposed to PFAS in contaminated sediments, which can disrupt their physiological functions, reproduction, and survival rates. For example, studies have shown that PFAS exposure reduces the reproductive success of benthic invertebrates such as mussels and amphipods, which are critical to maintaining healthy aquatic ecosystems (Cousins et al., 2022).

The transfer of PFAS from sediments to benthic organisms also facilitates bioaccumulation and biomagnification. As PFAS move up the food chain, they accumulate in higher concentrations in predatory species, ultimately impacting fish, birds, and even humans who rely on these ecosystems for sustenance. This process underscores the interconnectedness of sediment contamination and broader ecosystem health.

Furthermore, long-term sediment contamination has significant implications for water quality. Even when active sources of PFAS pollution, such as industrial discharges, are mitigated, sediments can act as secondary pollution sources, gradually releasing PFAS into the water over time. This chronic contamination hinders efforts to restore water bodies to safe conditions, necessitating sediment-focused remediation strategies. Techniques such as capping contaminated sediments, dredging, and innovative in situ treatments are being explored, but these methods are costly and pose additional risks of remobilizing pollutants during implementation (Grandjean et al., 2023).

Addressing PFAS in sediments requires a holistic approach that integrates sediment management with broader water quality improvement efforts. Effective remediation strategies must account for the complex interactions between sediments, water, and biological systems to break the cycle of contamination and protect both ecosystems and human health.

Emerging Concerns: PFAS in Polar and Remote Regions

One of the most alarming aspects of PFAS contamination is its infiltration into polar and remote regions, areas far removed from direct industrial activity. The global reach of PFAS is largely attributed to atmospheric transport, where volatile PFAS precursors, such as fluorotelomer alcohols (FTOHs), travel vast distances through air currents before depositing onto land or water surfaces via precipitation. This process, known as long-range atmospheric transport, has resulted in the widespread detection of PFAS in remote ecosystems, including the Arctic and Antarctic. Studies have shown significant PFAS concentrations in polar ice, snow, and water bodies, demonstrating how these "forever chemicals" transcend borders and industrial zones (Young et al., 2007; Cousins et al., 2019).

The presence of PFAS in polar regions poses unique challenges for local wildlife and ecosystems. Species such as seals, polar bears, and seabirds that are highly dependent on marine food webs are particularly vulnerable to PFAS bioaccumulation and biomagnification. For instance, apex predators like polar bears have been found to carry some of the highest PFAS burdens among wildlife due to their consumption of contaminated prey. This contamination not only impacts their health, manifesting in hormonal disruptions and immune suppression, but also jeopardizes the stability of polar ecosystems already under stress from climate change (Butt et al., 2010; Grandjean et al., 2023).

The interplay between PFAS contamination and climate change further exacerbates the situation in polar regions. As global temperatures rise and ice sheets melt, PFAS that were once sequestered in frozen environments are released into surrounding waters. This process introduces previously trapped contaminants into marine ecosystems, compounding existing pollution levels. Recent research has suggested that melting permafrost and glacial runoff are emerging sources of PFAS, creating a feedback loop where environmental change perpetuates chemical contamination (Liu et al., 2021).

Remote mountain regions are similarly affected by PFAS deposition, with contamination documented in alpine lakes and streams across the Himalayas, Alps, and Rockies. These ecosystems, often considered pristine, highlight the global scale of PFAS pollution and the pervasive nature of their environmental impact. For instance, PFAS deposition in alpine regions has been linked to atmospheric transport from urban and industrial areas, with precipitation serving as the primary pathway for chemical deposition. This has led to concerns about the long-term health of mountain ecosystems, including impacts on freshwater biodiversity and downstream water supplies (Nguyen et al., 2020).

The contamination of polar and remote regions underscores the interconnected nature of global environmental systems. It highlights the inability of any ecosystem, no matter how isolated, to remain untouched by human activity. Addressing PFAS in these regions requires not only localized mitigation strategies but also global efforts to reduce emissions and eliminate sources of PFAS at their origin.

The fight against PFAS contamination is not just a battle for clean water and healthy ecosystems, it is a decisive moment in humanity's responsibility to protect the planet. The extensive and lasting impacts of these chemicals, from their bioaccumulation in aquatic life to their infiltration of remote and fragile environments, underscore the gravity of the crisis we face. The damage to biodiversity, the contamination of critical water sources, and the disruption of ecological balance reminds us of the interconnectedness of all life on Earth.

As we grapple with the enduring consequences of PFAS, it is clear that half-measures will not suffice. The long-term contamination of sediments, the bioaccumulation within food webs, and the alarming spread of PFAS to polar and alpine regions demand immediate and unified global action. Addressing this challenge requires a multi-faceted approach, combining innovative scientific solutions, stringent regulatory frameworks, and a commitment to reducing reliance on these harmful substances.

The legacy of PFAS contamination serves as a cautionary tale, a stark reminder of the unintended consequences of technological progress when it outpaces environmental stewardship. Protecting the ecosystems impacted by PFAS is more than an environmental obligation; it is a moral duty to ensure the health and resilience of our planet for generations to come. In rising to meet this challenge, we affirm our commitment to safeguarding not only the Earth's ecosystems but also the very foundation of life itself.

Chapter 10

Advances in Detection and Monitoring

The detection and monitoring of PFAS have emerged as crucial areas of focus in the global effort to address these persistent environmental contaminants. With the ability of PFAS to remain in the environment for decades, accurately measuring their concentrations in surface waters is essential for understanding their distribution, behavior, and impacts. Advances in analytical techniques have significantly improved detection capabilities, but challenges remain in measuring ultra-trace levels of PFAS and harmonizing methods across regions. This chapter explores the latest developments in PFAS detection, the hurdles faced by researchers, and the importance of global collaboration in monitoring efforts.

Analytical Techniques for Measuring PFAS in Surface Waters

Advances in analytical chemistry have revolutionized the ability to detect and quantify PFAS in surface waters. Techniques such as liquid chromatography coupled with tandem mass spectrometry (LC-MS/MS) have become the gold standard for PFAS analysis due to their sensitivity and precision. LC-MS/MS allows for the simultaneous detection of multiple PFAS compounds at concentrations as low as parts per trillion (ppt), a significant improvement over earlier methods (Liu et al., 2021).

High-resolution mass spectrometry (HRMS) has further enhanced detection capabilities, enabling researchers to identify previously unknown PFAS compounds through non-targeted analysis. This approach is particularly valuable given the large number of PFAS variants in use and their potential to degrade into other forms over time (Nguyen et al., 2020). Additionally, the development of isotope-dilution techniques has improved quantification accuracy by accounting for matrix effects that can interfere with detection (Wang et al., 2017).

Recent innovations have also extended PFAS detection to environmental matrices beyond water, including sediments, soil, and biological tissues. These advancements provide a more comprehensive understanding of PFAS distribution and pathways, which is critical for effective monitoring and remediation strategies.

The 25 most tested PFAS compounds represent a fraction of the more than 12,000 PFAS chemicals in existence but are prioritized due to their widespread use, persistence, and documented health and environmental impacts (see table below). These compounds include legacy PFAS like perfluorooctanoic acid (PFOA) and perfluorooctanesulfonic acid (PFOS), which were historically used in nonstick cookware, firefighting foams, and water-resistant textiles. Short-chain replacements like perfluorobutanesulfonic acid (PFBS) and GenX chemicals have also come under scrutiny, as they are marketed as safer alternatives yet exhibit similar persistence and bioaccumulative properties (Kwiatkowski et al., 2020). Other frequently tested compounds include perfluorohexanesulfonic acid (PFHxS), perfluorononanoic acid (PFNA), and perfluorodecanoic acid (PFDA), which are commonly detected in drinking water, soil, and biological samples. Testing these compounds enables researchers to monitor the spread of PFAS contamination, assess exposure risks, and evaluate the effectiveness of remediation efforts. While these 25 compounds provide valuable data, they only scratch the surface of the PFAS problem, as emerging compounds and mixtures remain largely unregulated and poorly understood (Nguyen et al., 2020; Sims et al., 2025). Expanding analytical testing to include more PFAS compounds is critical for comprehensively addressing the global contamination crisis.

Compound Name	Common Uses/Origins
Perfluorooctanoic acid (PFOA)	e.g. Nonstick cookware
Perfluorooctanesulfonic acid (PFOS)	Fire foams, textiles, food packaging
Perfluorobutanesulfonic acid (PFBS)	Water-resistant products, fire foams
GenX (Hexafluoropropylene oxide dimer acid)	Teflon materials
Perfluorohexanesulfonic acid (PFHxS)	Waterproofing, firefighting foams
Perfluorononanoic acid (PFNA)	Plastics, food contact materials
Perfluorodecanoic acid (PFDA)	Surfactants, water repellents
Perfluoroundecanoic acid (PFUnDA)	Textile and paper treatments
Perfluorododecanoic acid (PFDoDA)	Industrial applications
Perfluorotridecanoic acid (PFTrDA)	Industrial and consumer products
Perfluorotetradecanoic acid (PFTeDA)	Long-chain surfactants
Perfluoropentanoic acid (PFPeA)	Industrial processing
Perfluorohexanoic acid (PFHxA)	Industrial and consumer products
Perfluoroheptanoic acid (PFHpA)	Industrial surfactants
Perfluorodecane sulfonic acid (PFDS)	Industrial applications
Perfluorooctane sulfonamide (PFOSA)	Legacy surfactant
N-ethyl perfluorooctane sulfonamidoethanol	Textiles and papers
N-methyl perfluorooctane sulfonamidoethanol	Textiles and papers
Perfluoro-3-methoxypropanoic acid (PFMPA)	Emerging contaminant
Perfluoro-2-methoxypropanoic acid (PFEtMPA)	Emerging contaminant
6:2 Fluorotelomer sulfonic acid (6:2 FTS)	Firefighting foams
8:2 Fluorotelomer sulfonic acid (8:2 FTS)	Firefighting foams
Perfluoro(2-ethoxyethane)sulfonic acid (PFEESA)	Water-resistant products
4:2 Fluorotelomer sulfonic acid (4:2 FTS)	Firefighting foams
Perfluoroisobutanoic acid (PFIBA)	Industrial/consumer applications

Challenges in Detecting Ultra-Trace Levels of PFAS

Despite significant advancements, detecting PFAS at ultra-trace levels remains one of the most demanding aspects of environmental monitoring. A major hurdle is the ubiquitous presence of PFAS in laboratory environments, including in analytical equipment, reagents, and even personal protective equipment used by analysts. These pervasive contaminants can lead to cross-contamination, false positives, and reduce confidence in results. Laboratories must implement stringent contamination control protocols, such as the use of PFAS-free materials, ultrapure solvents, and rigorous blank testing, to ensure data integrity. However, these measures are resource-intensive and may not be feasible for all facilities, particularly smaller or underfunded labs (Domingo et al., 2021).

The chemical diversity of PFAS compounds further complicates detection efforts. With over 12,000 variants identified, many of which have distinct physicochemical properties, developing methods that can detect a broad spectrum of PFAS at ultra-trace levels is a formidable task. For example, short-chain PFAS, often used as replacements for their long-chain counterparts, are highly mobile and exhibit reduced retention on analytical columns used in liquid chromatography-tandem mass spectrometry (LC-MS/MS). These properties make short-chain PFAS more challenging to separate and quantify compared to their long-chain analogs (Kwiatkowski et al., 2020).

Environmental matrices also play a critical role in complicating PFAS detection. Complex water samples, such as those from rivers, groundwater, or wastewater, often contain high levels of dissolved organic matter and other organic or inorganic substances that can interfere with PFAS signals. Such interferences can obscure low-concentration PFAS peaks, complicate quantification, and reduce recovery rates during sample extraction and preparation. Advanced sample preparation techniques, such as solid-phase extraction (SPE) coupled with isotope dilution, are often required but add to the cost and complexity of analysis.

Adding to these challenges, the ultra-low regulatory limits established by the U.S. Environmental Protection Agency (EPA) have pushed the boundaries of existing analytical capabilities. The EPA has set health advisory levels for PFAS in drinking water as low as parts per trillion (ppt), a threshold that challenges even state-of-the-art LC-MS/MS systems. Only the latest-generation instruments equipped with specialized add-ons, such as high-resolution mass analyzers, extended injection volumes, and optimized ionization sources, can reliably detect PFAS at these levels. Even with these advancements, achieving reproducibility at ultra-low concentrations requires high-volume injections, which are not feasible for every laboratory due to cost and throughput limitations.

These technical challenges underscore the need for continued innovation in PFAS detection. Research into alternative detection methods, such as time-of-flight mass spectrometry and electrochemical sensors, shows promise but requires further development and validation. To meet regulatory requirements and protect public health, investments in accessible, high-sensitivity detection technologies must be prioritized, alongside expanded training for laboratories worldwide.

Standardized Methods and Global Efforts in PFAS Monitoring

The lack of standardized methods for PFAS detection has been a persistent obstacle to effective global monitoring. Variations in protocols for sample collection, preparation, and analysis across laboratories and countries complicate efforts to compare data and evaluate the full extent of contamination. These inconsistencies hinder the ability to identify trends and develop comprehensive mitigation strategies. Recognizing this critical gap, organizations such as the U.S. Environmental Protection Agency (EPA) and the International Organization for Standardization (ISO) have begun developing standardized guidelines for PFAS detection in environmental samples. For example, EPA Method 537.1 focuses on measuring PFAS in drinking water, while ISO guidelines provide a framework for analyzing these chemicals in complex environmental matrices (U.S. EPA, 2022).

Global initiatives, such as the Global Monitoring Plan under the Stockholm Convention on Persistent Organic Pollutants, have sought to harmonize PFAS monitoring efforts across regions. These initiatives aim to establish regional monitoring networks, facilitate data sharing, and build capacity, particularly in developing countries where access to advanced analytical technologies is limited. Such programs help reduce disparities in PFAS monitoring capabilities and promote the development of consistent global datasets that are critical for assessing the scope of contamination (Grandjean & Clapp, 2015).

Technological advancements have played a crucial role in addressing challenges in PFAS monitoring. Among the most promising innovations is the growing use of direct injection methods in liquid chromatography-tandem mass spectrometry (LC-MS/MS). Unlike traditional techniques that require extensive sample preparation, direct injection allows for the analysis of PFAS with minimal handling, reducing the risk of contamination and improving accuracy. This approach is particularly valuable for detecting ultra-trace levels of PFAS, as it minimizes the loss of target compounds during preparation. Direct injection also shortens analysis times and increases sample throughput, making it a cost-effective and efficient solution for laboratories handling high volumes of samples (Sims et al., 2025). Additionally, advancements in LC-MS/MS instrumentation such as those made by Perkin Elmer (e.g. QSight 420) with specific add-ons, such as specialized ion sources and enhanced detector sensitivity, have made it possible to meet the ultra-low detection levels required by the EPA's stringent guidelines for PFAS in drinking water. These innovations are essential for providing actionable data in regions with high PFAS contamination.

Research collaborations have further advanced PFAS monitoring capabilities. The European PERFORCE3 project, for instance, unites scientists, policymakers, and industry stakeholders to refine analytical methods and improve the understanding of PFAS exposure pathways. Similarly, initiatives in the United States, such as the National PFAS Testing Strategy, aim to expand monitoring efforts and provide critical data to guide regulatory decisions. These collaborative approaches highlight the importance of pooling resources and expertise to tackle the global challenge of PFAS contamination (Cousins et al., 2019).

Despite these advancements, significant challenges remain. Developing standardized methods that account for the diversity of PFAS compounds, many of which exhibit unique chemical properties, requires continuous research and adaptation. Additionally, global monitoring efforts must address the inclusion of emerging PFAS compounds and ensure that low-resource regions have the tools and knowledge necessary to participate effectively. As technologies like direct injection and international collaborations continue to evolve, the prospect of a unified global strategy for PFAS monitoring becomes increasingly attainable.

Finally, advances in PFAS detection and monitoring are pivotal in addressing the global challenge posed by these persistent and harmful chemicals. Innovations in analytical techniques, such as LC-MS/MS, HRMS, and direct injection methods, have enhanced the ability to detect PFAS at increasingly lower concentrations, providing critical insights into contamination levels and exposure risks. However, significant challenges remain, including the complexity of measuring ultra-trace levels, managing the diversity of PFAS compounds, and harmonizing global monitoring standards. Addressing these obstacles demands sustained investment in cutting-edge research, the establishment of universally accepted standardized methods, and a commitment to international collaboration. By uniting scientific expertise, regulatory frameworks, and global resources, the path forward offers a powerful opportunity to mitigate PFAS contamination and protect both environmental and human health for generations to come.

Chapter 11

PFAS Toxicology

Per- and polyfluoroalkyl substances (PFAS) pose a significant toxicological concern due to their persistence, bioaccumulation, and widespread human exposure. Unlike many other environmental contaminants, PFAS resist metabolic breakdown and accumulate in tissues over time. Their toxic effects are multi-systemic, impacting the liver, endocrine system, immune function, reproductive health, and metabolism (Grandjean & Clapp, 2015).

Absorption, Distribution, Metabolism, and Excretion (ADME)

PFAS absorption occurs through multiple pathways, making human exposure nearly unavoidable in contaminated environments. The most common route is through drinking water, where PFAS-contaminated municipal supplies and private wells serve as primary sources of ingestion. Food consumption is another major exposure pathway, as PFAS accumulate in fish, meat, and dairy products due to their persistence in agricultural soils and water systems. Additionally, food packaging materials, particularly those designed to resist grease, can leach PFAS into food, further increasing dietary intake (Domingo et al., 2021). Inhalation also contributes to PFAS exposure, particularly in occupational settings where workers are exposed to airborne PFAS particles in manufacturing facilities, military sites, and wastewater treatment plants. Even in non-industrial environments, airborne PFAS from indoor dust, carpet treatments, and water-resistant textiles can enter the respiratory system and contribute to systemic accumulation (Nguyen et al., 2020). Dermal exposure is considered a less significant route of absorption compared to ingestion and inhalation, but PFAS have been detected in personal care products, waterproof clothing, and firefighting foams, raising concerns about skin absorption over prolonged periods. Once absorbed, PFAS quickly enter the bloodstream, where they distribute to various tissues and organs, leading to long-term health risks.

Once PFAS enter the bloodstream, they bind strongly to serum proteins, particularly albumin, facilitating their distribution to vital organs. This binding mechanism allows PFAS to accumulate in the liver, kidneys, and thyroid, organs that play critical roles in metabolism, detoxification, and hormonal regulation (Grandjean & Clapp, 2015). Unlike many other persistent organic pollutants, PFAS are not lipophilic in the traditional sense; however, certain compounds exhibit affinities for fatty tissues, prolonging their biological half-life and making their elimination from the body extremely slow. The widespread distribution of PFAS within the body has been linked to disruptions in metabolic processes, immune function, and endocrine signaling. Their ability to cross the placenta also poses significant risks to fetal development, as PFAS exposure during pregnancy has been associated with low birth weight, altered immune responses, and developmental delays in newborns. Due to their stability and strong protein-binding properties, PFAS remain in circulation for extended periods, increasing the likelihood of chronic health effects over time.

PFAS are characterized by their extreme resistance to metabolic breakdown, contributing to their persistence in the human body. Unlike other environmental contaminants that undergo enzymatic transformation in the liver, PFAS largely bypass metabolic degradation, allowing them to accumulate over time. Their elimination primarily occurs through the kidneys, but this process is highly inefficient, with reported human half-lives ranging from several years to decades depending on the specific compound (Post et al., 2012). Long-chain PFAS, such as perfluorooctanoic acid (PFOA) and perfluorooctanesulfonic acid (PFOS), exhibit particularly slow elimination rates due to their strong affinity for blood proteins and renal reabsorption mechanisms that prevent efficient excretion. Short-chain PFAS, while eliminated more quickly, still persist in the body long enough to cause concern, especially with repeated exposure. Breastfeeding serves as another elimination pathway, but this process raises additional concerns about infant exposure, as PFAS-contaminated breast milk can transfer these persistent chemicals to newborns (Domingo et al., 2021). The inefficient excretion and bioaccumulative nature of PFAS underscore the challenges in reducing human body burdens and highlight the urgent need for reducing environmental exposure at the source.

Toxicological Effects of PFAS

The toxicological effects of PFAS exposure are widespread, affecting multiple organ systems and physiological processes. As highly persistent and bioaccumulative chemicals, PFAS can cause long-term damage to the liver, endocrine system, reproductive health, immune function, and increase cancer risks. The growing body of scientific evidence underscores the urgency of addressing PFAS contamination to mitigate its impact on human health.

Hepatotoxicity (Liver Toxicity)

The liver is one of the primary organs affected by PFAS exposure. These chemicals have been associated with hepatomegaly (liver enlargement), enzyme abnormalities, and disruptions in lipid metabolism. Studies in exposed populations have shown that PFAS can elevate liver enzyme levels, indicating liver damage and dysfunction (Nguyen et al., 2020). Furthermore, PFAS exposure has been linked to increased cholesterol levels, contributing to cardiovascular risks even in individuals without prior metabolic disorders.

Animal studies have demonstrated that PFAS can induce non-alcoholic fatty liver disease (NAFLD), a condition characterized by excessive fat accumulation in the liver. Chronic exposure to PFAS can also lead to liver fibrosis, a precursor to more severe hepatic conditions such as cirrhosis and hepatocellular carcinoma (Grandjean & Clapp, 2015). The hepatotoxic effects of PFAS are particularly concerning given their persistence in the body and the liver's role in detoxification, further compounding the risks associated with prolonged exposure.

Endocrine and Metabolic Disruptions

PFAS are known endocrine disruptors, interfering with hormonal balance and metabolic regulation. One of the most well-documented effects is PFAS-induced thyroid dysfunction. These chemicals interfere with thyroid hormone production and transport, leading to increased risks of hypothyroidism, particularly in populations with long-term exposure (Domingo et al., 2021). Thyroid hormones play a crucial role in metabolism, growth, and development, making thyroid dysfunction a significant health concern, especially for pregnant women and children.

Beyond thyroid disruption, PFAS exposure has been linked to metabolic disorders, including insulin resistance, obesity, and type 2 diabetes. Research suggests that PFAS alter glucose metabolism and insulin signaling pathways, contributing to weight gain and an increased risk of metabolic syndrome (Nguyen et al., 2020). These effects have been observed in both human epidemiological studies and animal models, reinforcing concerns about the role of PFAS in the global rise of obesity and metabolic diseases.

Reproductive and Developmental Toxicity

PFAS exposure has been shown to negatively impact reproductive health in both men and women. Studies indicate that PFAS can reduce fertility by disrupting hormonal function, decreasing sperm quality in men, and impairing ovarian follicle development in women (Grandjean & Clapp, 2015). These disruptions can lead to difficulties in conceiving and an increased risk of reproductive disorders.

The developmental effects of PFAS exposure are equally concerning. Prenatal exposure to PFAS has been linked to low birth weight, preterm births, and developmental delays in children. Because PFAS cross the placental barrier, fetal exposure occurs in utero, potentially leading to long-term health issues such as altered immune function, endocrine disruption, and neurodevelopmental disorders (Domingo et al., 2021). Infants and young children are particularly vulnerable due to their developing organ systems, which are more susceptible to the toxic effects of PFAS.

Immunotoxicity and Cancer Risks

One of the most alarming aspects of PFAS exposure is its impact on the immune system. PFAS have been shown to suppress immune function, making individuals more susceptible to infections and reducing vaccine efficacy. Studies on children have found that higher PFAS blood levels correlate with lower antibody responses to routine vaccinations, such as those for tetanus and diphtheria (Grandjean & Clapp, 2015). This immunosuppressive effect raises concerns about population-wide vulnerability to infectious diseases, particularly in communities with high PFAS exposure.

PFAS exposure has also been strongly linked to an increased risk of cancer, particularly in highly exposed populations. Epidemiological studies have found strong associations between PFAS exposure and kidney, testicular, and liver cancers. The mechanisms behind PFAS-induced carcinogenesis include oxidative stress, DNA damage, and disruption of cell signaling pathways (Post et al., 2012). Industrial workers and communities living near PFAS-contaminated sites have shown significantly higher incidences of these cancers, emphasizing the need for stricter regulations and continued research into the long-term carcinogenic effects of PFAS.

Toxicity Thresholds and Risk Assessment

The establishment of toxicity thresholds for PFAS is a critical aspect of public health protection, guiding regulatory agencies in setting exposure limits and risk mitigation strategies. Due to the persistent and bioaccumulative nature of PFAS, even trace amounts in drinking water, food, or consumer products can lead to long-term health risks. As a result, regulatory agencies have taken increasingly stringent measures to define toxicity thresholds and establish advisory limits to protect public health.

Regulatory Guidelines and Ultra-Low Advisory Limits

Regulatory agencies worldwide have set exposure limits for PFAS, with the U.S. Environmental Protection Agency (EPA) leading efforts to impose stringent drinking water advisories. The EPA has recognized that PFAS pose significant health risks even at ultra-low concentrations, setting interim health advisory levels for perfluorooctanoic acid (PFOA) and perfluorooctane sulfonic acid (PFOS) as low as 0.004 parts per trillion (ppt) and 0.02 ppt, respectively (U.S. EPA, 2022). These advisory limits reflect the growing body of toxicological evidence suggesting that no safe level of exposure exists for some PFAS compounds, particularly given their cumulative effects over time.

International agencies have also responded to mounting concerns about PFAS toxicity. The European Chemicals Agency (ECHA) has proposed sweeping restrictions on PFAS in industrial applications, while individual nations such as Denmark and Germany have enacted stringent PFAS limits for drinking water and consumer products. Similarly, the World Health Organization (WHO) is working to align global drinking water standards with emerging toxicological data. These efforts highlight the increasing recognition of PFAS as a class of hazardous substances requiring urgent regulatory action (Grandjean & Clapp, 2015).

Toxicological Research and Risk Assessment

Toxicological research continues to play a critical role in refining risk assessments for PFAS exposure. While legacy compounds such as PFOA and PFOS have been extensively studied, newer and emerging PFAS variants, including short-chain alternatives, present significant regulatory challenges. Many short-chain PFAS, initially introduced as "safer" alternatives, exhibit similar persistence and mobility in the environment, raising concerns about their long-term health effects (Domingo et al., 2021).

One of the key challenges in PFAS risk assessment is determining cumulative exposure effects. Since PFAS are found in multiple environmental sources, including air, water, food, and consumer products, individuals may be exposed through multiple pathways simultaneously. Traditional risk assessments often focus on single-exposure sources, but emerging research suggests that aggregate exposure to PFAS mixtures could have additive or even synergistic toxic effects. This complexity underscores the need for more comprehensive epidemiological studies to evaluate long-term health risks.

Additionally, researchers are working to establish clearer dose-response relationships for PFAS toxicity. Unlike traditional contaminants, PFAS exhibit nonlinear toxicity, meaning that adverse effects may occur at very low concentrations without a clear threshold for harm. This phenomenon has been observed in studies examining PFAS-induced immune suppression, endocrine disruption, and cancer risks, further complicating regulatory decision-making (Nguyen et al., 2020).

Future Directions in PFAS Risk Assessment

As the scientific community gains a better understanding of PFAS toxicity, risk assessment methodologies are evolving to incorporate new analytical approaches. High-throughput screening methods, computational toxicology models, and biomonitoring studies are being used to predict toxicity profiles for lesser-known PFAS compounds. Additionally, advances in toxicokinetics, including better modeling of PFAS bioaccumulation and elimination rates, are improving risk characterization efforts (Post et al., 2012).

Regulatory agencies are also increasingly shifting toward a precautionary approach, moving away from compound-by-compound assessments toward class-based regulations. This paradigm shift acknowledges the challenges posed by the sheer number of PFAS chemicals and the impracticality of evaluating each individually. By implementing class-based restrictions, policymakers aim to prevent regrettable substitutions, where one hazardous PFAS is replaced by another with similar toxic properties (Kwiatkowski et al., 2020).

Ultimately, the future of PFAS risk assessment will rely on continued collaboration between scientists, regulators, and industry stakeholders. Strengthening global regulatory frameworks, investing in toxicological research, and expanding biomonitoring efforts will be essential to mitigating PFAS exposure risks and protecting public health.

Ultimately, the toxicology of PFAS reveals a chilling reality—these "forever chemicals" are not just persistent in the environment but within the very fabric of human biology, accumulating silently over time and wreaking havoc on vital organ systems. Their ability to disrupt endocrine function, compromise immune responses, and elevate cancer risks makes them one of the most insidious chemical threats of our era. Unlike other contaminants that degrade over time, PFAS defy nature's attempts at elimination, persisting across generations and magnifying their long-term health consequences. The urgency to act has never been greater. Strengthening regulations, advancing detection technologies, and prioritizing exposure reduction are not just scientific imperatives but moral obligations. Without decisive intervention, the toxic legacy of PFAS will continue to unfold, threatening not only present populations but the health of future generations. The time for action is now.

Chapter 12

Financial Costs of PFAS

Financial costs of PFAS contamination extend far beyond environmental and health impacts, posing significant economic burdens on governments, industries, and communities worldwide. From costly remediation efforts to lawsuits and economic losses in agriculture, fisheries, and tourism, PFAS exemplify the profound financial consequences of environmental mismanagement. As the global community grapples with these costs, the urgency for effective prevention, mitigation, and accountability has never been greater.

Economic Burden of PFAS Cleanup on Governments, Industries, and Communities

The cleanup of PFAS contamination represents one of the most expensive environmental challenges of modern times, with costs projected to rise as new contamination sites are discovered and regulatory requirements tighten. Governments at all levels are grappling with the staggering financial burden of testing, monitoring, and remediating contaminated water supplies and ecosystems. For example, in the United States alone, the Environmental Working Group (EWG) estimates that addressing PFAS contamination in drinking water systems will cost billions of dollars, potentially exceeding $20 billion in some scenarios, depending on the extent of contamination and the technologies implemented (EWG, 2022; Mahler et al., 2023).

Industries responsible for PFAS production or use, such as chemical manufacturers and firefighting foam producers, are increasingly being held accountable for cleanup costs through legal settlements and regulatory enforcement. For instance, 3M announced in 2023 a $10.3 billion settlement to resolve PFAS contamination claims related to public water systems across the U.S. (3M, 2023). However, the financial burden often shifts to local governments and taxpayers, particularly in communities that lack the resources to litigate against large corporations. Municipalities are forced to invest in advanced water treatment technologies, such as granular activated carbon (GAC), ion exchange resins, and reverse osmosis systems, which can cost millions of dollars to implement and maintain annually (Nguyen et al., 2020; EPA, 2022). Small water systems, in particular, face disproportionate challenges as they struggle to secure funding for these expensive solutions.

For affected communities, the economic impact extends beyond cleanup efforts. Residents face increased water bills to cover treatment costs, often leading to financial strain in already economically vulnerable populations. Reduced property values in contaminated areas compound these challenges, as homeowners find it difficult to sell properties impacted by PFAS contamination, eroding household wealth (Domingo et al., 2021). Additionally, lost productivity due to health-related issues linked to PFAS exposure adds to the economic toll, with some estimates suggesting that PFAS-related health costs in the U.S. could reach tens of billions of dollars annually due to increased cancer rates, immune system suppression, and other illnesses (Grandjean & Clapp, 2015).

These cascading financial burdens underscore the urgent need for equitable solutions that prioritize vulnerable populations and ensure that the costs of PFAS contamination do not fall disproportionately on those least able to afford them. Federal and state governments must provide financial assistance to affected communities while holding polluters accountable through stringent regulations and legal action. Without coordinated efforts, the economic impacts of PFAS will continue to grow, further exacerbating social and environmental inequities.

Costs of Remediation Technologies and Lawsuits

The remediation of PFAS-contaminated sites is an exceptionally costly and technically complex endeavor, posing significant financial challenges for governments, industries, and communities. Traditional remediation methods, such as pump-and-treat systems, involve extracting contaminated groundwater, treating it to remove PFAS, and reinjecting it into the aquifer. This process is both time-intensive and resource-demanding, often requiring decades of continuous operation. Advanced treatment technologies, including granular activated carbon (GAC), ion exchange resins, and high-pressure reverse osmosis (RO), are more effective but come with exorbitant costs. For instance, the installation of RO systems at municipal water treatment plants can cost upwards of $10 million, with annual operating expenses ranging from $500,000 to over $1 million, depending on the scale and water quality parameters (Nguyen et al., 2020; U.S. EPA, 2022).

Emerging technologies such as in situ chemical oxidation, electrochemical destruction, and plasma-based treatments hold promise for degrading PFAS directly in contaminated media. However, these approaches remain in experimental or pilot stages and are not yet scalable for widespread deployment. A recent study on plasma-based technology suggested it could achieve PFAS degradation in situ, but operational costs for a pilot setup were estimated at $5–$15 per cubic meter of treated water, highlighting the financial barriers to commercialization (Singh et al., 2023). Similarly, electrochemical oxidation, while promising, faces scalability challenges due to its high energy demands and potential byproducts, adding complexity to cost-effective implementation (Sharma et al., 2023).

Beyond cleanup expenses, legal settlements related to PFAS contamination have imposed substantial financial liabilities on companies responsible for producing or using PFAS. High-profile cases include the $671 million settlement by DuPont and Chemours in 2017 for contamination in the Ohio River Valley, a case that exposed widespread drinking water contamination and associated health risks (Bilott, 2019). Similarly, 3M announced a $10.3 billion settlement in 2023 to address PFAS contamination claims in U.S. public water systems, marking one of the largest environmental settlements in history (3M, 2023). However, while these settlements provide some financial relief to affected communities, they often fall short of covering the full cost of remediation and long-term health monitoring, leaving governments and taxpayers to shoulder the residual financial burden.

The cascading financial implications of PFAS contamination also affect public utilities and local economies. Municipal water systems face increased operational costs due to the need for advanced treatment technologies, while communities near contaminated sites often experience property devaluation and decreased economic activity in industries like real estate, tourism, and agriculture. The National Institute of Environmental Health Sciences (NIEHS) reported in 2022 that PFAS contamination in U.S. agriculture alone could result in billions of dollars in lost revenue due to soil and water contamination, highlighting the pervasive economic ripple effects of these chemicals (NIEHS, 2022).

Addressing the financial toll of PFAS contamination will require a multi-pronged approach, including stricter regulatory frameworks, increased federal and international funding for remediation research, and mandatory financial contributions from industries responsible for PFAS production and use. Without such measures, the economic burden of PFAS will continue to escalate, disproportionately affecting communities least equipped to bear these costs.

Financial Impact on Agriculture, Fisheries, and Tourism

PFAS contamination poses severe economic challenges for agriculture, fisheries, and tourism, disrupting livelihoods and imposing significant financial burdens on communities and industries. In the agricultural sector, PFAS-laden water and soils compromise crop production and livestock health. Farmers in contaminated regions face immense challenges, including decreased crop yields, the need to test for PFAS in food products, and the inability to market contaminated goods. For example, in Maine, several dairy farms were forced to shut down after PFAS were detected in milk and soil, resulting in millions of dollars in lost revenue and long-term economic hardship for affected families and communities (NIEHS, 2022). The cost of replacing contaminated water supplies for irrigation and adapting farming practices adds further strain, creating ripple effects throughout the agricultural supply chain.

The fisheries sector has also been significantly impacted by PFAS contamination, as these chemicals accumulate in fish and other aquatic species, rendering seafood unsafe for human consumption. Declining fish stocks due to contamination disrupt local ecosystems and devastate fishing communities that rely on these resources for their livelihoods. For instance, in Michigan, PFAS contamination in the Huron River led to a fishing advisory in 2018, significantly affecting local fishing industries and associated businesses (Michigan Department of Environment, Great Lakes, and Energy [EGLE], 2020). The need for regular testing of fish for PFAS levels further increases costs for fisheries, reducing profitability and market competitiveness.

Tourism, particularly in areas known for their natural beauty, is equally vulnerable to PFAS contamination. Polluted lakes, rivers, and coastal areas deter visitors, directly affecting local economies that depend on tourism revenue. Beach closures, fishing restrictions, and advisories for recreational waters caused by PFAS contamination leads to substantial financial losses for small businesses, hospitality industries, and regional tourism sectors. For example, in 2021, PFAS contamination in Lake Michigan prompted closures of popular recreational areas, leading to decreased tourist activity and reduced revenue for local businesses (Domingo et al., 2021). Such disruptions underscore the broad economic impacts of PFAS beyond direct environmental and public health consequences.

Addressing the economic fallout in these sectors requires immediate action, including federal and state-level funding for mitigation efforts, subsidies for affected industries, and investments in PFAS remediation technologies. Additionally, implementing stricter regulations to prevent further contamination is crucial to protect these vital economic sectors from ongoing financial instability.

The financial costs of PFAS contamination reveal the immense economic burden these "forever chemicals" impose on society, transcending environmental and health impacts to touch nearly every sector of the economy. From billions of dollars spent on remediation technologies and legal battles to the devastation of agriculture, fisheries, and tourism, PFAS contamination underscores the far-reaching consequences of unchecked pollution. For governments, industries, and communities, the path forward must combine innovation with accountability, developing cost-effective solutions while enforcing stringent regulations to prevent future contamination.

The economic case for addressing PFAS is undeniable: every dollar invested in proactive measures today can save countless resources tomorrow. Beyond financial considerations, tackling the PFAS crisis is a moral imperative to protect ecosystems, public health, and livelihoods. It is a call for decisive, collective action to secure a sustainable and equitable future for all.

Chapter 13

Regulatory Landscape

Regulation of per- and polyfluoroalkyl substances (PFAS) has been a contentious and evolving issue in the U.S. and around the world. While the health and environmental risks associated with these chemicals are widely acknowledged, regulatory responses have been slow, inconsistent, and often hindered by political and industrial resistance. This chapter explores the differences between U.S. regulations and global policies, the challenges in setting enforceable limits for PFAS, and the influence of lobbying efforts that have shaped, delayed, or weakened regulatory actions.

U.S. Regulatory Approach to PFAS: Fragmented and Reactive

The regulation of per- and polyfluoroalkyl substances (PFAS) in the United States has been largely fragmented, reactive, and inconsistent, with state governments often leading the charge in the absence of strong federal mandates (U.S. Environmental Protection Agency [EPA], 2023; Sunderland et al., 2019). Despite growing scientific evidence linking PFAS exposure to severe health risks, including cancer, endocrine disruption, immune system damage, and developmental disorder, federal regulatory actions have been slow, incomplete, and frequently influenced by industry lobbying (Grandjean & Clapp, 2015).

Historically, the EPA has focused on guidance and voluntary agreements rather than enforceable mandates, allowing chemical manufacturers and polluting industries to self-regulate (EPA, 2023). This approach has left millions of Americans exposed to PFAS-contaminated drinking water, soil, and air, particularly in areas near industrial facilities, military bases, and wastewater treatment plants (Hu et al., 2016).

Instead of proactively banning or restricting PFAS, the U.S. regulatory framework has relied on gradual, piecemeal interventions, resulting in prolonged public health risks and delayed environmental remediation efforts (Kwiatkowski et al., 2020).

Key Federal Actions and Shortcomings

The EPA's Health Advisory Levels (2016): A Weak First Step

In 2016, the EPA issued non-binding health advisory levels for PFOA and PFOS in drinking water, setting the threshold at 70 parts per trillion (ppt) for combined exposure to both chemicals (Grandjean & Clapp, 2015; Department of Defense, 2022; Hu et al., 2016; Kwiatkowski et al., 2020; Sunderland et al., 2019; USEPA, 2016; 2023). These advisories were intended to guide local water utilities in addressing PFAS contamination, but because they were not legally enforceable, they were largely ignored by many states and municipalities.

Why Was This a Problem?

- No legal consequences for exceeding the limits: Since the advisory levels were recommendations rather than enforceable regulations, many water utilities continued to supply contaminated water without consequence.

- Scientific evidence suggested the limits were too high: Independent studies indicated that even exposure below 70 ppt posed significant health risks, particularly for pregnant women, infants, and those with compromised immune systems.

- Failure to address the full spectrum of PFAS chemicals: The advisory levels only applied to PFOA and PFOS, while thousands of other PFAS compounds remained unregulated and continued to be used in manufacturing.

As public awareness of PFAS contamination grew, calls for stronger regulations intensified, eventually prompting federal action in the following years.

The 2021 EPA PFAS Roadmap: A Step in the Right Direction, But Too Slow

Under growing public and political pressure, the Biden administration released the PFAS Strategic Roadmap in 2021, laying out a plan to:

- Set enforceable drinking water limits for key PFAS chemicals.

- Designate certain PFAS compounds as hazardous substances under the Comprehensive Environmental Response, Compensation, and Liability Act (CERCLA), also known as the Superfund Law.

- Enhance environmental monitoring and research efforts to assess the extent of PFAS contamination.

- Strengthen public health protections and establish cleanup requirements for contaminated sites.

Challenges and Setbacks:

- Slow implementation: While the roadmap outlined an ambitious regulatory framework, actual policy enforcement has been delayed by bureaucratic hurdles and industry opposition.

- Industry pushback and legal challenges: Major chemical companies and trade associations have aggressively lobbied against stricter regulations, arguing that compliance costs would be excessive and that alternative PFAS compounds remain essential for manufacturing and national security.

- Limited funding for cleanup efforts: The Superfund designation would make polluters financially responsible for cleanup, but industry lawsuits and political debates have slowed this process.

Despite these setbacks, the roadmap signaled a shift toward stronger federal regulation, laying the groundwork for future mandatory PFAS restrictions.

The 2023 Proposed National Drinking Water Standards: A Long-Awaited Move

In 2023, the EPA proposed the first-ever enforceable drinking water standards for PFOA and PFOS, setting the limit at 4 ppt, a significant reduction from the previous advisory level of 70 ppt (USEPA, 2016; 2023). This move was widely praised by public health advocates and environmental organizations as a necessary step toward protecting communities from PFAS exposure.

Why Is This Important?

- Legally enforceable: Unlike the 2016 health advisory, these proposed standards would require water utilities to comply, ensuring safer drinking water for millions of Americans.

- Based on updated scientific evidence: The new limits reflect growing research showing that even trace amounts of PFAS can be harmful.

- Sets a precedent for broader PFAS regulation: This is the first time the EPA has attempted to regulate PFAS comprehensively at the federal level, paving the way for potential bans and manufacturing restrictions.

Remaining Challenges:

- Not yet finalized: The rules have not been fully implemented, and the regulatory process is still subject to legal and industry challenges.

- Industry resistance: PFAS manufacturers, water utilities, and industrial trade groups have pushed back, arguing that removing PFAS to such low levels will be prohibitively expensive.

- Limited scope: The proposed regulations only apply to PFOA and PFOS, leaving thousands of other PFAS compounds unregulated.

If fully enacted, these standards would force water utilities nationwide to upgrade their filtration systems and require stronger oversight of PFAS pollution. However, enforcement and industry compliance remain uncertain.

The Department of Defense (DoD) and PFAS: A History of Delayed Action

The U.S. military is one of the largest historical sources of PFAS contamination, primarily due to its extensive use of aqueous film-forming foam (AFFF) in firefighting training exercises (Department of Defense, 2022; USEPA, 2016; 2023). Military bases, airports, and training facilities have been identified as highly contaminated sites, exposing service members, surrounding communities, and groundwater supplies to high PFAS concentrations.

Key Issues:

- Massive contamination footprint: More than 700 military bases across the U.S. have been identified as PFAS-contaminated sites.

- Health impacts on veterans and base personnel: Service members exposed to PFAS through drinking water, soil, and firefighting foam training have reported higher rates of cancer, thyroid disease, and reproductive issues.

- Slow cleanup efforts: Despite Congressional pressure, the Department of Defense (DoD) has been slow to address contamination, often citing high cleanup costs and lack of available technology.

- Resistance to stricter regulations: The military has pushed back against proposed bans on PFAS-based firefighting foams, arguing that viable alternatives do not yet exist for critical fire suppression needs.

Recent Developments:

- In 2021, Congress mandated that the DoD phase out PFAS-containing firefighting foams by 2024, though implementation has been slow.

- In 2022, lawsuits were filed by military families and veterans seeking compensation for health impacts caused by PFAS exposure at military installations.

- The Pentagon has faced increasing scrutiny, but cleanup efforts remain underfunded and delayed, leaving many contaminated communities without immediate relief.

Despite growing awareness of PFAS contamination at military sites, the slow pace of cleanup efforts and resistance from the defense sector highlight the complex political and financial barriers to meaningful PFAS regulation.

Need for Stronger, Faster Federal Action

The U.S. regulatory approach to PFAS has been delayed, fragmented, and often shaped by industry influence. While recent efforts, including enforceable drinking water limits, regulatory roadmaps, and military phase-outs, represent progress, challenges remain. Public health risks continue to grow, and without stronger enforcement, broader chemical bans, and industry accountability, PFAS contamination will persist as a major environmental and public health crisis.

The coming years will determine whether the U.S. government can enact and enforce meaningful PFAS regulations, or if industry resistance and bureaucratic delays will keep Americans exposed to these toxic chemicals indefinitely.

State-Level Regulations: Filling the Federal Gap

As the federal government has struggled to implement and enforce comprehensive PFAS regulations, many individual states have taken matters into their own hands, enacting stricter laws to protect drinking water, food supplies, and consumer products (Department of Defense, 2022; Kwiatkowski et al., 2020; Sunderland et al., 2019; USEPA, 2016; 2023;). These state-level initiatives highlight the urgent need for action but also expose the inconsistencies in how PFAS contamination is managed across the country.

While some states have pioneered aggressive PFAS restrictions, others continue to lack enforceable standards, resulting in a patchwork of regulations that complicates compliance for businesses and creates disparities in public health protections.

Leading States in PFAS Regulation

Michigan: A Leader in PFAS Drinking Water Standards

Michigan has been at the forefront of PFAS regulation, setting some of the strictest enforceable drinking water limits in the country (Michigan Department of Environment, 2020; Michigan PFAS Action Response Team, n.d.; Troutman Pepper, 2020). In 2020, the state dramatically lowered its allowable limits for several PFAS compounds, including:

- PFOA: 8 ppt (compared to the EPA's proposed 4 ppt in 2023)

- PFOS: 16 ppt

- PFNA: 6 ppt

- GenX Chemicals: 370 ppt

Why This Matters

- Michigan has identified more than 200 PFAS-contaminated sites, many linked to industrial discharges and legacy pollution from firefighting foams and manufacturing waste.

- The state implemented aggressive testing requirements for public water supplies, identifying dozens of communities with high PFAS levels.

- PFAS cleanup efforts have been prioritized, with state-funded remediation programs targeting the most contaminated areas.

Despite these proactive steps, cleanup costs remain high, and polluters, including 3M and DuPont, continue to challenge state regulations in court.

California: Banning PFAS in Consumer Products

California has taken an aggressive approach to limiting PFAS exposure, focusing not just on drinking water but also on consumer products, food packaging, and industrial discharges (California Department of Toxic Substances Control, 2021; Intertek, 2024).

Key Actions Taken by California

- PFAS Ban in Food Packaging – California became one of the first states to prohibit PFAS in food contact materials, including fast-food wrappers, takeout containers, and microwave popcorn bags. This prevents PFAS from leaching into food and entering the human body through ingestion.

- Stricter Drinking Water Standards – The state requires monitoring and reporting of PFAS contamination in public water systems, with some of the strictest notification levels in the country.

- Consumer Product Restrictions – In 2022, California passed a law banning PFAS in cosmetics and personal care products, targeting waterproof mascaras, long-lasting foundations, and sunscreens.

- Firefighting Foam Restrictions – The use of PFAS-based firefighting foams has been heavily restricted, requiring fire departments to transition to PFAS-free alternatives by 2024.

California's tough stance on PFAS regulation has led to significant industry resistance, with companies arguing that alternative materials are expensive or less effective (California Department of Toxic Substances Control, 2021; Intertek, 2024). Despite these challenges, California continues to push for more comprehensive PFAS bans, setting a standard for other states to follow.

Maine: The First State to Ban All Non-Essential PFAS Products

Maine has gone even further than other states, enacting the most sweeping PFAS ban in the U.S (Maine Legislature, 2021). In 2021, Maine passed a law that will ban all PFAS-containing products by 2030, unless manufacturers can prove that their use is essential to public health and safety.

What Makes Maine's Law Unique?

- It targets all PFAS chemicals, not just PFOA and PFOS, ensuring that companies cannot replace phased-out PFAS with equally harmful alternatives.

- It shifts the burden of proof onto manufacturers, requiring them to demonstrate why PFAS use is necessary rather than assuming their safety.

- The law also mandates reporting requirements, forcing manufacturers to disclose their PFAS use to the state.

Maine's approach is being closely watched by other states and may serve as a model for future national policies (Maine Legislature, 2021). However, industry groups have pushed back aggressively, arguing that the ban is too broad and unrealistic for manufacturers to comply with.

New York & New Jersey: Toughest PFAS Water Quality Regulations

Both New York and New Jersey have enacted some of the strictest PFAS water regulations, setting legally enforceable limits that exceed federal guidelines (New Jersey Department of Environmental Protection, 2020; New York State Department of Environmental Conservation, 2016).

New York's Regulations

- In 2020, New York set a maximum contaminant level (MCL) of 10 ppt for both PFOA and PFOS—lower than the EPA's previous 70 ppt guideline.

- The state requires water utilities to monitor PFAS regularly and report contamination levels to the public.

- New York has allocated millions in state funding for PFAS remediation efforts in heavily impacted communities.

New Jersey's Regulations

- New Jersey was one of the first states to set binding limits for PFAS in drinking water, enforcing:

    - PFOA: 14 ppt

- PFOS: 13 ppt

- PFNA: 13 ppt

- The state has pursued legal action against major PFAS manufacturers, including 3M and DuPont, to force them to pay for environmental cleanup costs.

These strict PFAS regulations reflect growing concerns in the Northeast, where legacy contamination from manufacturing and industrial sites has heavily impacted groundwater and drinking water supplies.

Challenges of State-Level PFAS Regulation

State-level efforts have played a crucial role in advancing PFAS regulations, but the absence of a unified federal standard has created a patchwork of inconsistent policies. This lack of uniformity results in varying levels of protection across states. For instance, residents in states like New York and Michigan benefit from strict PFAS drinking water limits, while those in states without regulations remain vulnerable to contamination (New York State Department of Environmental Conservation [NYSDEC], 2020; Michigan Department of Environment, Great Lakes, and Energy [EGLE], 2020). Businesses also struggle with compliance as manufacturers, retailers, and water utilities must navigate different PFAS rules depending on the state, leading to increased costs and operational complexities (Interstate Technology & Regulatory Council [ITRC], 2023). Additionally, regulatory loopholes persist, as PFAS-contaminated products banned in one state can still be legally sold in another, prolonging public exposure to these harmful chemicals (Environmental Working Group [EWG], 2022).

Another major challenge is industry pushbacks and legal opposition. Chemical manufacturers and industry groups argue that state-specific PFAS regulations impose an unfair burden on businesses, particularly those operating in multiple states (Kwiatkowski et al., 2020).

Corporations like 3M and DuPont have filed lawsuits to challenge these regulations, aiming to delay or weaken their enforcement (U.S. Government Accountability Office [GAO], 2023). In some cases, economic concerns have pressured states to reconsider or relax their restrictions, especially in industries where PFAS use is widespread, such as aerospace, semiconductors, and pharmaceuticals (Sunderland et al., 2019).

Ultimately, the need for federal leadership is evident. Without a single, enforceable national standard, PFAS regulation will remain fragmented, leaving many communities without adequate protection (U.S. Environmental Protection Agency [EPA], 2023). While state-led initiatives have driven progress, a comprehensive federal response is necessary to establish uniform protections for all Americans, ensuring that no community is left vulnerable due to geographical disparities in regulation.

A Patchwork System in Need of Federal Action

State-level regulations have played a crucial role in filling the federal void, with Michigan, California, Maine, New York, and New Jersey setting some of the most stringent PFAS protections in the country (National Conference of State Legislatures [NCSL], 2023; U.S. Environmental Protection Agency [EPA], 2024a). These proactive measures have pressured the Environmental Protection Agency (EPA) to take more decisive federal action, highlighting the urgency of the PFAS crisis (EPA, 2024b). As a result, the Biden administration finalized the first-ever national drinking water standard for PFAS, an effort largely driven by state leadership in addressing contamination (EPA, 2024c; JD Supra, 2024). Without these aggressive state actions, federal policymakers may have continued to delay intervention, leaving many communities vulnerable to ongoing PFAS exposure (EPA, 2023).

However, a fragmented system of state-by-state regulations is not a long-term solution. Without a strong, enforceable national standard,

PFAS contamination will continue to disproportionately impact communities based on their state's regulatory priorities.

To truly address the PFAS crisis, the federal government must:

- Establish strict nationwide drinking water limits.

- Ban non-essential PFAS use in consumer products.

- Hold manufacturers accountable for contamination and cleanup.

- Provide funding for water treatment and environmental remediation.

Until these actions are taken, state governments will continue to bear the burden of protecting their residents from the widespread and persistent threat of PFAS contamination.

Global PFAS Policies: A Comparative Perspective

While the United States has struggled to implement enforceable PFAS regulations, several other countries and international organizations have taken a more proactive approach. Unlike the reactive stance seen in the U.S., many nations have adopted a precautionary principle, banning or restricting PFAS even before full scientific consensus on their health effects is reached. These regulatory measures vary widely, from strict bans on PFAS in consumer products and drinking water limits to legal actions against manufacturers.

This section explores key global efforts in regulating PFAS, highlighting how different regions have responded to the environmental and public health threats posed by these persistent chemicals.

European Union (EU): Leading the Charge in PFAS Regulation

The European Union (EU) has been at the forefront of PFAS regulation, taking one of the strictest stances globally. The EU's regulatory philosophy follows the precautionary principle, meaning that potential risks are addressed before significant harm is confirmed—a stark contrast to the reactive approach of the U.S.

Key Actions Taken by the EU

The Stockholm Convention on Persistent Organic Pollutants (POPs) – PFOA Ban (2020)

- In 2020, the EU banned perfluorooctanoic acid (PFOA) and its related compounds under the Stockholm Convention on Persistent Organic Pollutants (POPs)

- This ban prohibited the manufacture, use, and import of PFOA, recognizing its extreme persistence, bioaccumulation, and toxicity.

- The ban extended to a wide range of consumer and industrial applications, including nonstick cookware, textiles, firefighting foams, and food packaging.

The 2023 Proposed EU-Wide PFAS Ban

- In 2023, the European Chemicals Agency (ECHA) proposed a sweeping ban on all PFAS chemicals, one of the most ambitious regulatory efforts worldwide

- If fully enacted, this proposal would phase out thousands of PFAS compounds across manufacturing, consumer goods, and industrial applications over the next decade.

- The ban is supported by multiple EU member states, including Denmark, Germany, the Netherlands, Sweden, and Norway, all of which have pushed for aggressive PFAS restrictions

Why Is the EU's Approach Significant?

- The EU has focused on preventing PFAS pollution before it happens, rather than waiting for cleanup solutions.

- The ban covers entire classes of PFAS chemicals, unlike U.S. regulations, which focus only on a handful of specific compounds.

- The proposal includes strict liability for manufacturers, making industries financially responsible for remediation and damages caused by PFAS pollution.

Scandinavian Leadership: Denmark, Sweden, and Norway

The Nordic countries, Denmark, Sweden, and Norway, have long been pioneers in environmental and chemical regulation, and their PFAS policies are among the strictest in the world. These nations have taken aggressive steps to limit PFAS use in food packaging, drinking water, and consumer products.

In 2020, Denmark became the first country to ban per- and polyfluoroalkyl substances (PFAS) in paper and cardboard food contact materials, including items like fast-food wrappers, pizza boxes, and microwave popcorn bags. This legislation, effective from July 1, 2020, prohibits the use of PFAS-treated grease-resistant coatings in such products. Denmark has also advocated for the European Union to adopt similar bans, setting a precedent for other nations (Danish Ministry of Environment and Food, 2020).

Sweden has taken legal action against companies responsible for PFAS-contaminated drinking water, seeking compensation for public health damages and environmental cleanup costs. In a landmark ruling, the Swedish Supreme Court favored residents affected by PFAS contamination, entitling them to compensation (ChemSec, 2023). Additionally, Swedish regulatory agencies require strict monitoring and reporting of PFAS contamination in water supplies. The Swedish government is also pushing for PFAS to be classified as hazardous under EU chemical laws, making polluters financially responsible (ECHA, 2023).

Norway mandates that companies report the use of PFAS in consumer goods, including textiles, electronics, and industrial applications. The government has invested in PFAS-free alternatives and provides funding for research into non-toxic substitutes. Norway, along with other countries, has proposed a comprehensive restriction on PFAS under the EU's REACH regulation, aiming to limit their manufacture and use due to associated health and environmental risks (ECHA, 2023).

These Scandinavian nations continue to advocate for stronger EU-wide regulations and are leading voices in the international fight against PFAS pollution.

Canada and Australia have made notable strides in addressing per- and polyfluoroalkyl substances (PFAS) contamination, yet their regulatory frameworks exhibit inconsistencies across provinces and states.

Canada: Drinking Water Limits and Regulatory Gaps

Health Canada has established a drinking water objective of 30 nanograms per liter (ng/L) for the sum of 25 PFAS compounds, reflecting a commitment to stringent water quality standards (Health Canada, 2023). However, the implementation of these guidelines varies by province. For instance, while provinces like Quebec and British Columbia have introduced tougher PFAS restrictions, others have lagged behind, leading to a patchwork of regulations nationwide (Osler, Hoskin & Harcourt LLP, 2022). Moreover, unlike the European Union, Canada has not enacted comprehensive PFAS bans, allowing many consumer products containing these substances to remain on the market (Osler, Hoskin & Harcourt LLP, 2022).

Australia: Firefighting Foam Bans and Emerging Consumer Regulations

Australia has taken significant steps by banning PFAS-based firefighting foams, particularly in military, aviation, and industrial applications (Corrs Chambers Westgarth, 2021). For example, New South Wales prohibited the use of PFAS-containing firefighting foam for all training and demonstration purposes from April 2021, with further restrictions implemented in September 2022 (Corrs Chambers Westgarth, 2021). Similarly, South Australia has completely banned the use of fluorinated foams (Oil Technics, n.d.). Despite these advancements, PFAS regulations in consumer products are still developing, and the use of these chemicals persists in various industries. Additionally, PFAS laws vary by state, creating inconsistencies in regulation and enforcement across the country (Pinsent Masons, 2023).

While both Canada and Australia have made progress in limiting PFAS, their regulations remain less comprehensive than those in the European Union, and enforcement is not yet uniform across all provinces and states.

China: A Key Producer with Weak Regulations

China stands as one of the largest producers of per- and polyfluoroalkyl substances (PFAS), supplying these chemicals to global markets. However, its domestic regulations on PFAS use and pollution remain less stringent compared to regions like the European Union and the United States (CIRS Group, 2022).

Key Issues with China's PFAS Policies

Unlike the EU and U.S., China has not implemented comprehensive bans on PFAS, allowing their ongoing use in industries such as textiles, electronics, and various industrial applications (Antea Group, 2022). Some Chinese factories have become suppliers of PFAS chemicals that have been phased out elsewhere, exporting them to countries where their use is still permitted (Envirotech Online, 2023). The Chinese government has expressed intentions to tighten environmental regulations concerning PFAS, with plans to prohibit the import and export of perfluorooctane sulfonic acid (PFOS) products starting in 2024, though broader regulatory progress has been gradual and inconsistent (CIRS Group, 2022).

Global Pressure on China

As more countries enact PFAS bans, China may face pressure to align with stricter international standards to maintain access to global markets (Enhesa, 2022). Additionally, international trade agreements and a growing consumer demand for PFAS-free products could incentivize Chinese manufacturers to reduce or eliminate PFAS in their exports (Roland Berger, 2023). Despite its current regulatory stance, China's significant role as a leading PFAS producer means that its policies will have global implications, influencing the future availability and trade of PFAS-containing products.

Conclusion: Global Trends in PFAS Regulation

While the United States has been criticized for a slow and reactive approach to PFAS regulation, other nations and international organizations have taken more proactive measures to curb PFAS use and exposure.

Key Takeaways:

- The European Union is at the forefront of PFAS regulation, with proposals for comprehensive bans that could significantly impact global markets (International Chemical Regulatory and Law Review, 2023).

- Scandinavian countries, including Denmark, Sweden, and Norway, have implemented stringent bans on PFAS in food packaging and consumer goods and have pursued legal actions against polluters (Enhesa, 2022).

- Canada and Australia have made advancements in PFAS regulation, but enforcement varies across regions, leading to inconsistencies (Osler, Hoskin & Harcourt LLP, 2022).

- China, as a major PFAS producer, currently maintains less stringent regulations, though international pressure could prompt stricter policies (CIRS Group, 2022).

The global movement toward PFAS elimination is gaining momentum, compelling companies and governments to adapt to stricter regulations, develop safer alternatives, and address the legacy of contamination. The trajectory of U.S. policy in this context remains to be seen, but the international trend is clearly toward more rigorous control of PFAS substances.

Challenges in Setting Enforceable Limits for PFAS

Regulating per- and polyfluoroalkyl substances (PFAS) has proven to be one of the most difficult environmental and public health challenges of modern times. While scientific evidence increasingly confirms the dangers of PFAS exposure, governments worldwide have struggled to establish clear, enforceable limits for these chemicals (U.S. Environmental Protection Agency [EPA], 2023).

Several factors contribute to this challenge, including the complexity of PFAS chemistry, scientific uncertainty, high remediation costs, legal opposition from powerful corporations, and industry lobbying efforts to delay regulation (Enhesa, 2022).

Despite these hurdles, public pressure, legal actions, and international momentum are gradually pushing regulatory agencies to act (International Chemical Regulatory and Law Review, 2023). Below, we explore the major obstacles to establishing strong PFAS limits and enforcement mechanisms.

Complexity of PFAS Chemistry

One of the greatest obstacles to effective PFAS regulation is the sheer number and diversity of PFAS compounds. PFAS, or per- and polyfluoroalkyl substances, represent a vast family of synthetic chemicals with thousands of different structures, each designed for specific industrial and consumer uses. These compounds vary widely in chain length, functional groups, and chemical properties, which influences their environmental persistence, mobility, and potential toxicity. This structural diversity makes it incredibly challenging to create uniform regulatory standards or cleanup guidelines, as each compound can behave differently in water, soil, air, and biological systems. Additionally, scientific understanding of the environmental fate, bioaccumulation, and long-term health effects of many PFAS compounds is still evolving, further complicating efforts to manage and mitigate their risks.

Thousands of PFAS, Each with Unique Risks

- There are more than 12,000 known PFAS chemicals, each with different structures, properties, and toxicity levels (EPA, 2023).

- Some PFAS, such as perfluorooctanoic acid (PFOA) and perfluorooctane sulfonic acid (PFOS), have been extensively studied and linked to serious health issues, leading to their phase-out in many countries (European Chemicals Agency [ECHA], 2023).

- However, many newer "short chain" PFAS alternatives, such as GenX and Perfluorobutane sulfonic acid (PFBS), remain in widespread use, and their long-term health effects are still poorly understood (Agency for Toxic Substances and Disease Registry [ATSDR], 2022).

Regulatory Complexity

- Unlike traditional pollutants (e.g., lead, mercury, or asbestos), PFAS chemicals do not break down naturally and persist in the environment for decades or even centuries (ECHA, 2023).

- Developing a single, enforceable standard that applies to all PFAS is extremely difficult because different PFAS compounds behave differently in water, soil, and air (ATSDR, 2022).

- Regulators must decide whether to set limits for individual PFAS chemicals— which is time-consuming and allows industry to replace banned chemicals with slightly modified versions—or ban entire PFAS classes, a move opposed by manufacturers claiming that some PFAS are less harmful (Enhesa, 2022).

These challenges make it difficult for regulatory agencies like the EPA, EU authorities, and national governments to establish comprehensive PFAS policies that are both enforceable and scientifically sound.

Scientific Uncertainty and Industry Pushback

Scientific evidence has established a strong link between per- and polyfluoroalkyl substances (PFAS) exposure and severe health risks, including cancer, liver damage, hormone disruption, and immune system suppression (Agency for Toxic Substances and Disease Registry [ATSDR], 2022). Despite this, industry groups have argued that low doses are not harmful or that there is insufficient scientific data to justify strict regulation.

Deliberate Industry Strategy: "Doubt is Our Product"

- Chemical manufacturers, including 3M and DuPont, have been accused of deliberately downplaying health risks, employing tactics similar to those used by the tobacco and fossil fuel industries—casting doubt on scientific findings to delay regulation (Grandjean & Clapp, 2015).

- Internal company documents from lawsuits have revealed that 3M and DuPont were aware of the health dangers of PFAS as early as the 1970s but failed to disclose this information to regulators or the public (Grandjean & Clapp, 2015).

- Industry-backed studies have attempted to contradict independent research, suggesting that PFAS exposure at low levels is safe, despite overwhelming evidence to the contrary (Grandjean & Clapp, 2015).

The Problem of "Safe" Exposure Levels

- There is no universally agreed-upon "safe" level of PFAS exposure because even extremely low doses have been linked to negative health effects (ATSDR, 2022).

- Some studies suggest there is no safe threshold at all, making the establishment of enforceable limits both controversial and politically challenging (ATSDR, 2022).

As a result, regulatory agencies often delay setting strict limits, citing the need for further research, while industries continue to profit from PFAS production and use.

Cost of Compliance and Remediation

One of the primary arguments against strict PFAS regulations comes from industries and local governments, which claim that compliance with stringent limits would be prohibitively expensive.

High Costs of Water Treatment and Cleanup

- PFAS chemicals are notoriously difficult to remove from water supplies, requiring advanced and costly filtration systems such as granular activated carbon (GAC), ion exchange resins, and reverse osmosis membranes (U.S. Environmental Protection Agency [EPA], 2023).

- Upgrading public water treatment plants to remove PFAS could cost billions of dollars nationwide, leading to resistance from local governments and utility companies (EPA, 2023).

- Industries that have contaminated water supplies with PFAS oppose strict regulations because they could be held financially responsible for costly cleanup efforts (EPA, 2023).

Opposition from Businesses Using PFAS

- Industries such as aerospace, textiles, food packaging, cosmetics, and electronics rely on PFAS for properties like waterproofing, grease resistance, and durability (National Institute for Occupational Safety and Health [NIOSH], 2023).

- Many of these businesses argue that PFAS alternatives are not yet available or cost-effective, leading them to oppose outright bans and instead push for more lenient regulatory measures (NIOSH, 2023).

The economic burden of PFAS cleanup and compliance continues to be a significant barrier to establishing enforceable federal and global limits.

Legal Challenges and Corporate Resistance

Chemical manufacturers, particularly 3M and DuPont, have aggressively opposed per- and polyfluoroalkyl substances (PFAS) regulations through various legal and corporate strategies.

- **Litigation Against Regulatory Bodies**: These companies have filed lawsuits against the Environmental Protection Agency (EPA) and state governments to block or delay new PFAS regulations. For instance, 3M sued the state of Michigan in 2021, seeking to invalidate its PFAS drinking water standards (Ellison, 2021).

- **Avoiding Liability**: Corporations have utilized legal loopholes to evade responsibility for PFAS contamination. DuPont, for example, has restructured its corporate entities, attempting to shift liability to newly formed subsidiaries (Lerner, 2018).

- **Settlements Without Admission of Guilt**: Both 3M and DuPont have settled numerous lawsuits related to PFAS contamination without admitting fault. In 2023, 3M agreed to a $10.3 billion settlement to resolve claims of PFAS pollution in public water systems (3M, 2023).

Corporate Evasion Strategies

- **"Regrettable Substitution"**: When specific PFAS chemicals, such as PFOA, are banned, companies often replace them with structurally similar compounds like GenX. These substitutes may be equally harmful but lack comprehensive studies, allowing continued use under different names (Blake, 2018).

- **Shifting Legal Responsibility**: Some corporations have restructured or spun off subsidiaries to avoid liability for PFAS pollution. DuPont's creation of Chemours is a notable example, where liabilities for PFAS contamination were transferred to the new entity (Lerner, 2018).

Despite mounting lawsuits and legal settlements, corporate resistance continues to stall enforcement efforts at the federal level.

The Influence of Lobbying and Political Barriers

The chemical industry has significantly influenced PFAS policy through lobbying, public relations campaigns, and political donations aimed at delaying and weakening regulations.

Key Industry Tactics

- **Lobbying Against Strict PFAS Laws**: From 2019 to 2022, major PFAS manufacturers and their trade groups spent over $110 million lobbying Congress to weaken proposed federal regulations (Food & Water Watch, 2023). Industry representatives have argued that stringent PFAS drinking water limits would impose excessive cleanup costs (The New Lede, 2023).

- **Funding Doubtful Science**: Corporations have financed studies that downplay the risks of PFAS, creating regulatory uncertainty and slowing the adoption of stricter policies. This strategy mirrors tactics used by the tobacco industry to cast doubt on scientific evidence (Pulitzer Center, 2023).

- **Litigation and Delays**: PFAS manufacturers have challenged EPA and state-level rules in court, stalling enforcement and delaying protective measures for communities. For example, industry groups have filed lawsuits to contest new drinking water standards for PFAS (Reuters, 2024).

Despite these barriers, public awareness, legal action, and international pressure are compelling regulatory agencies to act.

The Growing Pressure for Change

- **Legal Accountability**: States have initiated lawsuits against PFAS manufacturers to hold them accountable for environmental contamination. As of December 2024, thirty-one state attorneys general have filed litigation related to PFAS pollution (Safer States, 2024).

- **Scientific Research**: Ongoing studies continue to reveal the health risks associated with PFAS exposure, undermining industry-funded research that suggests safety at low levels (ATSDR, 2022).

- **International Influence**: The European Union's stringent regulations on PFAS are prompting global companies to reassess their use of these chemicals, potentially influencing U.S. policy (ECHA, 2023).

Activists, scientists, and affected communities persist in advocating for stronger protections, keeping PFAS regulation in the global spotlight. The critical question remains whether governments will confront industry pressure to enforce meaningful limits that protect public health or allow PFAS contamination to continue unabated.

The Future of PFAS Regulation: A Defining Moment

PFAS regulation stands at a crossroads, with governments worldwide facing a stark choice: take decisive action to mitigate the damage or allow industrial influence to continue delaying critical reforms. While recent policy shifts signal progress, enforcement remains uneven, corporate accountability is still lacking, and the burden of contamination disproportionately falls on vulnerable communities. The next decade will determine whether policymakers have the resolve to close these gaps or if regulatory inertia will allow PFAS pollution to persist unchecked.

One undeniable truth remains, without bold and immediate intervention, PFAS contamination will not only endure but intensify, leaving future generations to bear the cost of inaction. The time to act is now. Will we rise to the challenge, or will we leave behind an irreversible legacy of environmental and public health devastation? The answer will define the future of PFAS regulation, and the health of generations to come.

Chapter 14

The Role of Industry

The widespread contamination of per- and polyfluoroalkyl substances (PFAS) is not a natural occurrence but a direct consequence of industrial production and application. PFAS have been used in countless consumer and industrial products for decades due to their water- and grease-resistant properties. However, their persistence in the environment and severe health risks have made them a regulatory priority (Agency for Toxic Substances and Disease Registry [ATSDR], 2022). As the public becomes increasingly aware of PFAS contamination, industries that manufacture and rely on these chemicals are under mounting pressure to transition toward safer alternatives. However, corporate accountability, legal challenges, and economic interests continue to slow this transition.

This chapter explores the industries most responsible for PFAS pollution, the corporate response to growing regulations, and potential pathways for industries to phase out PFAS while maintaining functionality in their products. Key Industries Responsible for PFAS Pollution

Per- and polyfluoroalkyl substances (PFAS) have been detected in everything from household dust to drinking water, but a handful of industries account for the vast majority of contamination. These industries have historically benefited from the unique chemical properties of PFAS, prioritizing performance, durability, and cost-effectiveness over environmental concerns. However, as the risks of PFAS become more evident, these industries are facing increasing scrutiny and regulatory pressure to transition to safer alternatives.

Firefighting Foam and Military Applications

One of the most significant contributors to PFAS pollution has been aqueous film-forming foam (AFFF), which has been widely used for fire suppression, particularly in military, aviation, and industrial settings. AFFF was designed to quickly smother liquid fuel fires, making it essential for high-risk environments such as airports, oil refineries, and military bases (United States Environmental Protection Agency [EPA], 2023).

However, the same properties that make AFFF effective in firefighting also make it a long-lasting environmental hazard. AFFF contains high concentrations of PFAS, which leach into groundwater and persist for decades (Hu et al., 2016). Numerous military installations and surrounding communities have reported widespread PFAS contamination in drinking water, leading to health concerns and costly remediation efforts (Andrews & Naidenko, 2020).

Despite mounting evidence of harm, regulatory efforts to phase out AFFF have been met with resistance. The aviation industry, for instance, has argued that non-PFAS firefighting foams do not yet meet the same performance standards, raising concerns about potential risks to safety (European Chemicals Agency [ECHA], 2023). While alternatives such as fluorine-free foams (F3) are emerging, adoption has been slow, and industry groups continue to push for extended timelines before full phase-outs (EPA, 2023).

In response to growing public and legal pressure, some governments have taken action. In 2020, the U.S. Congress mandated the phase-out of PFAS-based firefighting foams by 2024 for military installations under the National Defense Authorization Act (NDAA) (U.S. Department of Defense, 2021). Meanwhile, the European Union has proposed a total ban on AFFF containing PFAS, though implementation remains under debate due to industry concerns about fire safety efficacy (ECHA, 2023).

Textiles and Apparel

The textile and apparel industry are the largest users of PFAS, employing these chemicals in waterproof, stain-resistant, and durable finishes for products such as:

- Outdoor gear (raincoats, hiking boots, and gloves)

- Upholstered furniture and carpets

- Stain-resistant clothing and uniforms

PFAS have been particularly popular in high-performance apparel due to their ability to repel both water and oil while maintaining breathability. Major brands like Gore-Tex, Patagonia, and The North Face have historically used PFAS in their outerwear, though growing consumer awareness and regulatory pressure have forced some companies to commit to phasing them out (Blum et al., 2021).

However, the shift away from PFAS in textiles remains slow and inconsistent. Many companies argue that non-PFAS alternatives do not offer the same level of performance or durability, particularly for extreme weather and high-performance applications. Some brands, while pledging to remove PFAS, have simply switched to short-chain PFAS alternatives, which are less studied but may still pose significant health risks (Kwiatkowski et al., 2020).

Regulatory efforts targeting PFAS in textiles are gaining traction. In 2022, California banned PFAS in most textiles by 2025, requiring manufacturers to disclose PFAS usage and provide safer alternatives (California State Legislature, 2022). The European Union has proposed banning all non-essential PFAS in consumer textiles, though this has been met with pushback from textile manufacturers who claim that PFAS-free alternatives are not yet scalable (ECHA, 2023).

As regulatory frameworks tighten, the textile industry will need to accelerate the development and adoption of PFAS-free materials. New technologies, such as silicone-based and plant-derived water-repellent coatings, are being explored as safer alternatives, but widespread implementation remains a challenge due to cost concerns and production limitations (Blum et al., 2021).

Food Packaging and Consumer Goods

PFAS are widely used in food packaging and consumer goods due to their nonstick, water-resistant, and grease-resistant properties. Common PFAS-containing products include:

- Fast-food wrappers, takeout containers, and pizza boxes

- Microwave popcorn bags and disposable plates

- Non-stick cookware (Teflon and similar coatings)

- Toothpaste, dental floss, and personal care products

The food industry's dependence on PFAS has led to concerns about food contamination, as PFAS can leach from packaging into food, increasing human exposure (Schaider et al., 2017). Studies have found PFAS residues in fast food and supermarket products, raising alarms about the long-term health risks associated with dietary PFAS exposure (Blum et al., 2021).

Due to mounting evidence, some governments have taken action:

- In 2020, Denmark became the first country to ban all PFAS in food packaging (Danish Ministry of Environment, 2020).

- Several U.S. states, including California, New York, and Washington, have passed laws banning PFAS in food packaging (Environmental Working Group [EWG], 2022).

- The EU is considering a bloc-wide ban on PFAS in all food contact materials, but industry pushbacks have slowed progress (ECHA, 2023).

Despite these efforts, enforcement remains inconsistent, and PFAS-laden packaging continues to be widely available in regions where bans have not yet been implemented. Some food companies have voluntarily phased out PFAS, but many have replaced them with alternative coatings whose long-term safety is still unknown (Blum et al., 2021).

Industries such as firefighting, textiles, and food packaging have heavily relied on PFAS for decades, leading to widespread environmental contamination and human exposure. While some companies are beginning to transition to PFAS-free alternatives, many still resist change, citing performance, durability, and economic concerns. Regulatory efforts are gaining momentum, but corporate resistance, legal loopholes, and inconsistent enforcement continue to hinder progress.

As governments and consumers demand stronger protections against PFAS, industries will need to accelerate their transition to safer alternatives. The challenge now lies in ensuring that replacements are truly safe and do not repeat the mistakes of the past.

Corporate Accountability and Resistance to Regulation

Despite overwhelming scientific evidence linking per- and polyfluoroalkyl substances (PFAS) to severe health effects, many corporations have actively resisted regulation. Chemical manufacturers, including 3M, DuPont, and Chemours, have employed tactics similar to those used by tobacco, fossil fuel, and pharmaceutical industries to delay accountability. These tactics include misleading the public, suppressing internal data, challenging regulations through legal maneuvers, and lobbying policymakers to weaken laws (Grandjean & Clapp, 2015).

Corporate resistance has significantly slowed regulatory progress, allowing the continued production and use of PFAS under different names and formulations. The industry's ability to evade responsibility while profiting from harmful chemicals raises serious ethical and environmental concerns, especially as PFAS contamination continues to spread globally.

Deliberate Misinformation and Legal Evasion

Internal company documents, obtained through lawsuits and investigative journalism, reveal that 3M and DuPont were aware of the health risks of PFAS exposure as early as the 1970s but actively concealed this information (Grandjean & Clapp, 2015).

- Corporate Cover-Ups: Internal memos show that company scientists at 3M and DuPont identified PFAS-related health risks in laboratory animals and human studies decades ago. However, instead of disclosing their findings, these corporations kept the data confidential and continued mass production (Pulitzer Center, 2023).

- Funding Doubtful Science: In a deliberate effort to undermine regulatory efforts, 3M and DuPont funded studies that downplayed PFAS toxicity, much like the tobacco industry did to challenge smoking regulations (Pulitzer Center, 2023). These industry-sponsored studies introduced doubt into the scientific and regulatory process, slowing down policy interventions and making it more difficult for governments to set enforceable PFAS limits.

- Shifting Blame: As regulatory scrutiny intensified, DuPont restructured its corporate structure, spinning off its fluoropolymer division into a new company, Chemours, in

2015. This move effectively shifted liability for PFAS contamination away from DuPont and onto a separate entity, minimizing legal exposure while continuing the production of PFAS under different trade names (Lerner, 2018).

- Rebranding PFAS: Another key corporate evasion tactic has been "regrettable substitution," where one banned PFAS compound (e.g., PFOA) is replaced with a chemically similar but less-studied alternative (e.g., GenX). GenX was introduced as a "safer" alternative, but research now suggests it may be just as toxic as the PFAS it replaced (Blake, 2018).

Legal Challenges and Settlements

While 3M, DuPont, and Chemours have faced numerous lawsuits and agreed to multi-billion-dollar settlements, these legal outcomes often come without admissions of wrongdoing, allowing these corporations to continue producing and selling alternative PFAS compounds (Reuters, 2024).

- Multi-Billion-Dollar Settlements: In 2023, 3M agreed to a $10.3 billion settlement to resolve claims of PFAS contamination in public water systems (Reuters, 2024). DuPont, Chemours, and Corteva collectively agreed to a $1.2 billion settlement, but these companies did not admit liability and continue to sell PFAS-based products (EWG, 2023).

- Ongoing Litigation: Despite these settlements, lawsuits from states, municipalities, and affected communities continue to mount. As of 2024, dozens of states have filed legal action against PFAS manufacturers, arguing that the settlements do not adequately cover the long-term costs of PFAS remediation (Safer States, 2024).

- Corporate Denial Strategies: Even as they settle lawsuits, PFAS manufacturers have publicly denied the full extent of their responsibility, often claiming that their newer PFAS formulations are safer, despite growing evidence to the contrary (Blake, 2018).

Industry Lobbying and Policy Influence

The chemical industry has spent millions lobbying Congress and regulatory agencies to weaken PFAS legislation, delay regulatory enforcement, and limit corporate liability (Food & Water Watch, 2023).

Key Industry Tactics

Lobbying Against Strict PFAS Laws

- From 2019 to 2022, PFAS manufacturers spent over $110 million lobbying U.S. lawmakers (Food & Water Watch, 2023).

- Industry representatives have argued that stringent PFAS drinking water standards would impose excessive costs on businesses, municipalities, and water utilities, shifting the financial burden onto consumers (The New Lede, 2023).

Funding Industry-Friendly Research

- Chemical manufacturers have funded studies that downplay the risks of PFAS exposure, creating regulatory uncertainty and delaying the adoption of stricter environmental policies (Pulitzer Center, 2023).

- By highlighting gaps in toxicity data, industry-backed researchers have argued that more studies are needed before enforcing stricter PFAS limits, despite independent research linking even low-level exposure to serious health risks (ATSDR, 2022).

Challenging Regulations in Court

- PFAS manufacturers have repeatedly sued the EPA and state governments, arguing that PFAS regulations are too broad, too costly, or lacking in scientific certainty (Reuters, 2024).

- These lawsuits have stalled enforcement efforts, allowing corporations to continue PFAS production while legal battles unfold (EWG, 2023).

Corporate accountability for PFAS contamination has been systematically undermined by misinformation, legal evasion, and aggressive lobbying efforts. Despite clear evidence linking PFAS to environmental and health crises, chemical manufacturers continue to downplay risks, shift liability, and resist regulation.

While lawsuits and regulatory actions have forced some level of corporate accountability, PFAS manufacturers remain financially powerful and politically influential, continuing to shape policies in their favor. The fight against PFAS contamination is not just an environmental issue, it is a battle against corporate interests prioritizing profits over public health.

As scientific research and public awareness grow, the demand for corporate responsibility and stricter regulations will only intensify. The question remains: Will governments stand up to industry pressure and enforce meaningful change, or will PFAS contamination persist for future generations to bear the consequences?

Transitioning to PFAS-Free Industrial Practices

Industries that have historically depended on per- and polyfluoroalkyl substances (PFAS) are now facing mounting consumer pressure, stricter regulations, and increasing scientific evidence linking PFAS to serious health and environmental consequences. As a result, many companies are actively seeking PFAS-free solutions that maintain performance while reducing environmental harm.

However, the transition away from PFAS is not uniform across industries, with some sectors making more progress than others. While viable alternatives exist, challenges related to cost, technological limitations, and regulatory inconsistencies continue to slow widespread adoption.

Current Innovations and PFAS-Free Alternatives

Numerous industries are developing and testing PFAS-free materials that aim to provide the same durability, nonstick properties, and resistance to heat, water, and chemicals. Some of the most notable advancements include:

Firefighting Foam Alternatives

Firefighting foams have been one of the most significant sources of PFAS contamination, particularly at military bases, airports, and industrial sites. The primary challenge has been finding alternative foams that meet strict fire safety standards without sacrificing effectiveness.

- Fluorine-Free Foams (F3): Some military bases and commercial airports have begun testing PFAS-free foams, known as fluorine-free foams (F3), which have demonstrated promising results in controlled fire suppression tests (EPA, 2023).

- Europe Leading the Way: Countries such as Norway and Sweden have already banned PFAS-based firefighting foams and successfully implemented fluorine-free alternatives in commercial and industrial fire suppression systems (ECHA, 2023).

- Challenges: Some aviation and military organizations have pushed back against the phase-out of PFAS foams, arguing that current alternatives do not provide the same fire-extinguishing speed or durability in high-risk environments (EPA, 2023). However, ongoing research aims to close this performance gap and enhance F3 technology.

Textile Industry Innovations

The textile industry has relied on PFAS for waterproofing, stain resistance, and durability for decades. However, due to increasing consumer demand for eco-friendly materials, several brands are investing in PFAS-free textile technologies.

- Natural Fiber Blends: Some companies are turning to wax-coated cotton, wool, and plant-based fibers as sustainable alternatives to PFAS-treated fabrics (Blum et al., 2021).

- Silicone-Based Coatings: Silicone-based waterproofing treatments are gaining popularity as a durable and non-toxic alternative for outdoor apparel (Kannan et al., 2022).

- Leading Brands: Major brands like Patagonia, The North Face, and Gore-Tex are developing PFAS-free water-repellent materials, though adoption remains slow due to cost concerns and performance variability (Blum et al., 2021).

Food Packaging Alternatives

PFAS have been widely used in fast-food wrappers, microwave popcorn bags, and disposable plates due to their grease-resistant properties. However, as the risks of PFAS leaching into food become more widely recognized, alternative materials are gaining traction.

- PFAS-Free Biodegradable Coatings: Some food companies have transitioned to bio-based coatings made from cornstarch, cellulose, and other plant-derived materials that offer similar grease-resistant properties (Schaider et al., 2017).

- Danish Ban on PFAS in Food Packaging: Denmark was the first country to ban PFAS in food contact materials, leading other European nations to consider similar restrictions (Danish Ministry of Environment, 2020).

- Challenges in Adoption: While safer alternatives exist, many food companies hesitate to transition due to higher production costs and the need for new supply chain infrastructure (ECHA, 2023).

Non-Stick Cookware and Consumer Goods

The cookware industry has long relied on Teflon (polytetrafluoroethylene, or PTFE), a PFAS-based nonstick coating. However, concerns over PFAS leaching into food have driven demand for PFAS-free alternatives.

- Ceramic-Coated Cookware: Many cookware brands have moved toward ceramic coatings, which provide nonstick properties without PFAS exposure risks (Kwiatkowski et al., 2020).

- Cast Iron and Stainless Steel: Traditional cast iron and stainless-steel cookware are experiencing a resurgence as consumers seek durable, non-toxic alternatives to PFAS-coated products.

- Challenges: While ceramic coatings and stainless-steel cookware are gaining popularity, some consumers still prefer Teflon for its superior nonstick properties, making the transition gradual (Blum et al., 2021).

Challenges in Transitioning

While PFAS-free alternatives are emerging, industries face several barriers to full-scale adoption, including cost concerns, technological hurdles, and regulatory inconsistencies.

High Costs of Transitioning

- Research and Development Costs: Developing new PFAS-free materials requires extensive scientific research, testing, and certification, leading to higher costs for manufacturers (ECHA, 2023).

- Infrastructure Upgrades: Companies may need to redesign production lines and re-train employees, further increasing transition expenses (Blum et al., 2021).

- Pricing Competition: PFAS-containing materials are cheap and widely available, making it difficult for PFAS-free alternatives to compete on price without government subsidies or incentives (Schaider et al., 2017).

Performance and Durability Concerns

- Some PFAS-free alternatives do not yet match the performance of PFAS-based materials, particularly in industries like firefighting, outdoor apparel, and aerospace (EPA, 2023).

- Manufacturers worry that transitioning to less effective alternatives could lead to consumer dissatisfaction and product failure, especially in high-risk applications (ECHA, 2023).

Regulatory Inconsistencies

- While some countries and states have banned PFAS in certain products, there is no universal global standard, making compliance complex for multinational corporations (Blake, 2018).

- Some industries continue to exploit regulatory gaps, using short-chain PFAS as substitutes, despite growing evidence that they pose similar health risks (Grandjean & Clapp, 2015).

The transition to PFAS-free industrial practices is no longer an option, it is a necessity. The scientific evidence is undeniable, the environmental toll is staggering, and the public health risks are too severe to ignore. Yet, despite mounting pressure, progress remains frustratingly slow, hampered by corporate resistance, regulatory loopholes, and the high cost of change.

While firefighting foams, textiles, food packaging, and cookware are seeing promising advancements, incremental progress is not enough. A full-scale phase-out of PFAS requires an urgent and unified effort—from policymakers enforcing aggressive regulations, to industries prioritizing safer alternatives, to consumers demanding accountability from corporations.

The reality is clear: If we fail to act decisively, PFAS contamination will continue to spread, compounding the damage for future generations. The time for half-measures is over. Governments must lead with bold legislation, industries must innovate with purpose, and consumers must refuse to support companies that put profits over public health.

This is a defining moment in environmental and public health policy. The question is no longer whether we can eliminate PFAS, but whether we will act fast enough to prevent irreversible harm. The future is still unwritten—but our choices today will determine whether PFAS remains an enduring threat, or a cautionary tale of corporate greed finally brought to justice.

Chapter 15

Media and Public Awareness

Public awareness of per- and polyfluoroalkyl substances (PFAS) has grown significantly in recent years, largely due to investigative journalism, media coverage, and grassroots advocacy efforts. Once an issue confined to scientific research and environmental policy, PFAS contamination has now entered mainstream public discourse, prompting widespread concern about its health risks, regulatory failures, and corporate accountability.

The media plays a critical role in shaping how the public perceives PFAS, from exposing corporate misconduct to disseminating scientific findings. However, public perception is often influenced by misinformation, industry-backed narratives, and the complexity of PFAS science, which can create confusion and slow regulatory action. Meanwhile, grassroots organizations and affected communities have been instrumental in pushing for policy changes, corporate accountability, and stricter PFAS regulations.

This chapter explores the role of investigative journalism in exposing PFAS pollution, the challenges of public perception and misinformation, and the impact of grassroots advocacy in driving change.

Role of Investigative Journalism and Media in Shaping the PFAS Conversation

Investigative journalism has been one of the most powerful forces in exposing the truth about PFAS contamination. For decades, corporations like 3M and DuPont concealed evidence of PFAS-related health risks, and government agencies were slow to regulate these chemicals (Grandjean & Clapp, 2015). However, journalists have played a crucial role in bringing these issues to light and holding industries accountable.

One of the most high-profile examples of PFAS investigative journalism was The New York Times Magazine's 2016 exposé, "The Lawyer Who Became DuPont's Worst Nightmare", which detailed attorney Rob Bilott's legal battle against DuPont and its contamination of drinking water in Parkersburg, West Virginia (Rich, 2016). This investigation, along with others from media outlets like The Intercept, The Guardian, and The Washington Post, revealed that DuPont had known for decades that PFAS chemicals were hazardous to human health but continued to use them in consumer products without disclosure (Lerner, 2018).

The impact of investigative reports on PFAS contamination has been profound, significantly shaping public awareness, policy decisions, and corporate accountability. Mainstream media coverage has played a crucial role in increasing public knowledge about the dangers of PFAS exposure. Once again, documentaries such as *The Devil We Know* (2018) and the feature film *Dark Waters* (2019), which dramatized attorney Rob Bilott's legal battle against DuPont, helped bring PFAS concerns to a global audience, making the issue more tangible and pressing for everyday citizens (DuPont, 2023). As a result of these revelations, governments and regulatory agencies have faced mounting pressure to act. In response to growing public outcry, the U.S. Environmental Protection Agency (EPA) proposed its first-ever enforceable PFAS drinking water limits in 2023, marking a critical step in addressing widespread contamination (EPA, 2023).

Beyond raising awareness and influencing policy, investigative journalism has also played a significant role in holding corporations accountable. Reports exposing the long-standing deception of chemical manufacturers have led to multi-billion-dollar settlements, forcing companies like 3M and DuPont to acknowledge their role in environmental contamination (Reuters, 2024). These legal actions have helped pave the way for stricter regulations and corporate responsibility, though many argue that these settlements do not go far enough in addressing the long-term impact of PFAS pollution.

Despite these successes, media coverage of PFAS is not without challenges. Industry lobbying efforts, strategic misinformation campaigns, and the inherent complexity of PFAS science have all influenced how these issues are reported and perceived. Corporations continue to fund research that downplays PFAS risks, casting doubt on independent scientific findings and delaying regulatory action. Additionally, the slow-moving nature of PFAS contamination, which often takes years to manifest in measurable health effects, makes it harder to sustain media attention compared to more immediate environmental crises. As a result, while investigative journalism has been instrumental in exposing the dangers of PFAS, ongoing public engagement and media vigilance are essential to ensure continued progress in addressing this widespread issue.

Public Perception of PFAS Risks and Misinformation

Public understanding of per- and polyfluoroalkyl substances (PFAS) risks is shaped by a complex interplay of scientific research, media coverage, corporate narratives, and government messaging. While some communities, particularly those directly affected by PFAS-contaminated water supplies, are highly aware of the dangers, misinformation and uncertainty continue to hinder widespread public action. The long-term, invisible nature of PFAS contamination makes it a uniquely challenging issue to communicate effectively, as many people struggle to grasp the urgency of a threat that does not have immediate, visible consequences.

One of the greatest obstacles in public perception is the complexity of PFAS science. Unlike oil spills or air pollution, PFAS contamination is not immediately apparent and does not have a distinct smell, color, or immediate health effect. Instead, PFAS slowly accumulate in the body and environment over time, making their long-term impacts difficult for the average person to conceptualize (Kwiatkowski et al., 2020). This delayed effect weakens the sense of urgency and allows corporations and policymakers to justify gradual, rather than immediate, regulatory action. As a result, even when scientific studies link PFAS to severe health issues, public pressure for change often falls short of what is needed for swift regulatory intervention.

Adding to the confusion, PFAS manufacturers have actively engaged in misinformation campaigns, much like the tobacco and fossil fuel industries before them. Chemical corporations have funded studies designed to downplay the risks of PFAS exposure, leading to contradictory reports that create doubt and slow down regulatory action (Pulitzer Center, 2023). By leveraging public relations strategies and funding industry-friendly research, these companies have successfully discredited independent scientific findings and delayed stricter regulations, all while continuing to produce and sell PFAS under different trade names.

Further complicating public perception is the lack of clear safety standards. Unlike lead or mercury, which have well-established toxic thresholds, there is no universally accepted "safe" level of PFAS exposure. The Environmental Protection Agency (EPA) has proposed limits, but some scientific studies suggest that even trace amounts of PFAS can cause adverse health effects, including cancer and endocrine disruption (Grandjean & Clapp, 2015). This discrepancy leaves consumers uncertain about whom to trust, particularly when government agencies, industry representatives, and independent researchers offer conflicting guidance on what levels of PFAS are truly hazardous.

In an age where social media plays a dominant role in shaping public discourse, misinformation about PFAS spreads quickly. Some advocacy groups and scientists use Twitter, Facebook, and Instagram to educate the public, while anti-regulation groups and industry-funded accounts spread misleading claims about the necessity and safety of PFAS (Food & Water Watch, 2023). This digital battleground makes it increasingly difficult for the average consumer to discern fact from corporate propaganda, further delaying decisive action against PFAS contamination.

Ultimately, the lack of clear, consistent messaging on PFAS risks has led to public confusion and slowed regulatory responses. While investigative journalism and grassroots activism have helped raise awareness, the scientific complexity of PFAS, industry-backed misinformation, and regulatory uncertainty continue to pose significant barriers to widespread public mobilization. For meaningful progress to occur, governments, media outlets, and scientists must work together to counter misinformation, simplify PFAS science for the public, and establish clearer, enforceable safety standards that prioritize human and environmental health over corporate interests.

Role of Social Media in Public Awareness

Social media has become a powerful tool in shaping public awareness of per- and polyfluoroalkyl substances (PFAS), both as a means of spreading accurate information and as a platform for misinformation. Advocacy groups, environmental organizations, and independent researchers have used Twitter, Facebook, Instagram, and TikTok to educate the public about the dangers of PFAS contamination. Through infographics, viral posts, and investigative threads, these groups have raised awareness, debunked industry myths, and mobilized communities to demand regulatory action (Food & Water Watch, 2023).

However, social media has also been exploited by anti-regulation groups, industry-funded organizations, and chemical manufacturers, who have used the same platforms to spread misleading claims about PFAS safety. These efforts often mirror tactics used by the tobacco and fossil fuel industries, casting doubt on scientific findings and suggesting that PFAS bans would harm economic growth and innovation. Industry-backed narratives frequently downplay health risks, overstate the cost of regulation, and suggest that alternatives to PFAS are inadequate, creating confusion among consumers and policymakers. This digital battleground has polarized the conversation, making it harder for the public to separate scientific fact from corporate propaganda.

Despite these challenges, social media has played an essential role in amplifying grassroots activism, exposing corporate deception, and pressuring policymakers to take action. Digital campaigns have helped elevate the voices of affected communities, generate petitions for stricter PFAS regulations, and demand transparency from corporations. By leveraging the reach and immediacy of social media, activists have kept PFAS contamination in the public spotlight, preventing industry and government agencies from ignoring or downplaying the issue.

How Grassroots Advocacy Drives Change

While government agencies and corporations have been slow to address PFAS contamination, grassroots activism has become one of the most effective forces in driving regulatory action and holding polluters accountable. Community-led efforts, legal battles, and persistent advocacy have not only increased public awareness but also forced corporations to take responsibility and policymakers to enact stricter regulations.

Community-Led Activism and Legal Action

One of the most powerful aspects of grassroots advocacy is the direct action taken by affected communities. In regions where drinking water contamination has been confirmed, residents have mobilized to demand clean water, hold polluters accountable, and push for stronger PFAS regulations. These actions include public protests, petitions, town hall meetings, and legal challenges (Safer States, 2024).

Legal action has also been a critical tool for grassroots activists, leading to historic settlements that have forced corporations to pay for the damage caused by decades of PFAS pollution. In 2023, 3M and DuPont reached a $10.3 billion settlement to address PFAS contamination in U.S. drinking water systems (Reuters, 2024). While these settlements represent a step toward corporate accountability, many activists argue that they fall short of fully addressing the long-term environmental and health impacts of PFAS exposure.

At the state level, grassroots activism has successfully influenced policy changes, demonstrating that local action can create ripple effects on national regulations. Activists in Michigan, New York, and California have successfully pushed for state bans on PFAS in consumer products, setting important legal precedents that other states are beginning to follow (Environmental Working Group [EWG], 2023). These state-level victories have been instrumental in pressuring federal agencies to consider broader PFAS regulations, proving that persistent community-led efforts can shape national policy.

As PFAS contamination continues to affect communities worldwide, grassroots activism remains a driving force in demanding justice, pushing for stronger policies, and holding corporations accountable. By organizing at the local, state, and national levels, activists have shown that public pressure can influence regulatory decisions, force corporate transparency, and ensure that the voices of those impacted by PFAS pollution are heard.

Non-Profit Organizations and Public Advocacy

Beyond investigative journalism and grassroots activism, non-profit organizations have played a pivotal role in the fight against per- and polyfluoroalkyl substances (PFAS) contamination. These organizations have worked tirelessly to educate the public, pressure governments, and support independent scientific research aimed at finding alternatives to PFAS. Groups such as Earthjustice, the Environmental Working Group (EWG), and the PFAS Project Lab have been at the forefront of this battle, ensuring that corporate accountability and environmental justice remain central issues in PFAS policy discussions.

Education has been a key focus of these organizations. By publishing reports, hosting public forums, and engaging with the media, they have disseminated critical information about the health risks of PFAS exposure, helping to counter industry-backed misinformation. Through policy advocacy and legal action, these groups have also petitioned federal and state governments to adopt stricter PFAS regulations. Their efforts have led to bans on PFAS in food packaging, firefighting foams, and textiles in certain states, demonstrating the power of sustained public advocacy.

Additionally, non-profits have played a crucial role in advancing scientific research on PFAS alternatives. By funding independent studies and supporting university-led research initiatives, these organizations have helped accelerate the development of safer, PFAS-free materials. Their work has not only informed policymakers but has also provided viable solutions for industries transitioning away from PFAS-based products.

The fight against PFAS contamination is not just a scientific or regulatory issue, it is a battle for public awareness, corporate accountability, and environmental justice. Investigative journalism has been instrumental in exposing decades of corporate deception, while grassroots movements have mobilized communities and pushed for stronger regulations. However, the challenges remain significant. Industry-backed misinformation, the complexity of PFAS science, and the slow pace of regulatory change continue to obstruct meaningful progress.

For real change to occur, governments must prioritize transparency, enforce stricter PFAS regulations, and hold corporations accountable. Media outlets must continue to shine a light on industry misconduct, ensuring that PFAS contamination remains a public priority. Most importantly, the public must remain informed, engaged, and willing to demand action. The fight against PFAS is far from over, but as history has shown, public pressure and relentless advocacy have the power to transform policy, protect public health, and force industries to act responsibly. The question is not whether PFAS will be regulated, but whether we will act swiftly enough to prevent further harm. The time for action is now.

Chapter 16

PFAS and Climate Change

The connection between per- and polyfluoroalkyl substances (PFAS) contamination and climate change is an emerging area of concern in environmental science. As the planet warms, the effects of climate change are intensifying PFAS contamination, making it more difficult to mitigate and regulate these persistent chemicals. PFAS, often called "forever chemicals" due to their extreme resistance to degradation, have already been detected in drinking water, soil, air, and even human bloodstreams (Grandjean & Clapp, 2015). Climate change adds another layer of complexity by altering how PFAS move through the environment, increasing exposure risks, and complicating cleanup efforts.

Extreme weather events, such as hurricanes, floods, and wildfires, are becoming more frequent and severe due to climate change. These disasters disrupt PFAS-contaminated sites, mobilizing these chemicals and spreading them into new areas. Rising temperatures, changing precipitation patterns, and prolonged droughts also affect water resources, challenging existing strategies for PFAS removal from drinking water supplies. As governments and regulatory agencies work to control PFAS contamination, they must now account for climate-driven environmental changes that threaten to worsen the problem.

This chapter explores how climate change exacerbates PFAS contamination, the interactions between extreme weather events and PFAS pollution, and the challenge of protecting water resources in a warming world.

How Climate Change Exacerbates PFAS Contamination

Climate change is fundamentally altering the natural processes that govern the movement, persistence, and impact of PFAS in the environment. As global temperatures rise, the hydrological cycle, atmospheric patterns, and soil chemistry are shifting, influencing how PFAS are transported, accumulated, and ultimately ingested by humans and wildlife. These changes are making it increasingly difficult to control PFAS pollution, turning what was once a localized contamination issue into a widespread, global crisis.

One of the most concerning factors is the impact of rising global temperatures on water cycles, air currents, and soil composition, which significantly influences how PFAS spread and accumulate. PFAS are known for their chemical stability and resistance to environmental degradation, allowing them to persist for decades or even centuries. However, with climate change accelerating atmospheric and hydrological shifts, PFAS are now being redistributed in ways that make their containment even more challenging.

Higher temperatures increase evaporation rates, enhance atmospheric transport, and alter precipitation cycles, allowing PFAS to travel far beyond their original sources of contamination. Studies suggest that volatile PFAS compounds, such as fluorotelomer alcohols (FTOHs), can undergo long-range atmospheric transport, meaning they can evaporate, travel through the air, and later settle in remote regions where they accumulate in water, soil, and biological systems (Cousins et al., 2022). This phenomenon explains why PFAS have been detected in Arctic ice, rainwater, and the bodies of polar wildlife, even in areas with no known industrial PFAS sources. The process of global atmospheric redistribution of PFAS means that these chemicals are not just a local or regional problem, they are now a persistent and growing global contaminant.

The changing precipitation patterns driven by climate change also play a crucial role in PFAS mobility. As extreme rainfall events become more frequent, increased surface runoff washes PFAS from contaminated soils into groundwater, rivers, and lakes, further polluting essential drinking water sources. In urban environments, stormwater systems are often overwhelmed by heavy rainfall, increasing the risk of untreated PFAS-contaminated water being discharged into local waterways (Wee & Aris, 2023).

Conversely, regions experiencing prolonged droughts face an entirely different PFAS-related challenge (Sims et al., 2025). Water scarcity forces communities to rely on shrinking reservoirs, aquifers, and other water sources, where PFAS concentrations become more concentrated due to the reduced volume of available water (Halaly, 2025). This intensifies human exposure to PFAS, as people are consuming and using higher concentrations of contaminated water in their daily lives. Drought conditions also disrupt the natural dilution and dispersion mechanisms that normally help to mitigate the impact of PFAS in aquatic systems, further compounding the contamination crisis.

Another critical factor influenced by climate change is soil composition and microbial activity, which affects the breakdown and movement of PFAS in the environment. Higher temperatures can accelerate the degradation of certain PFAS precursors, leading to the formation of more stable, persistent PFAS compounds that remain in the environment for longer periods (Cousins et al., 2022). In addition, increased wildfires—another consequence of climate change, alter soil chemistry, reducing its ability to filter and retain PFAS, thereby allowing these chemicals to seep more easily into groundwater.

As these climate-related changes continue, managing PFAS contamination will become increasingly difficult, requiring new strategies for water treatment, pollution control, and environmental remediation. Without intervention, climate change will only accelerate the spread of PFAS, increasing their impact on human health, ecosystems, and critical water resources worldwide.

Interactions with Extreme Weather Events and PFAS Pollution

Extreme weather events fueled by climate change are exacerbating PFAS pollution, spreading contamination further, increasing human and ecological exposure, and making cleanup efforts more complex. As global temperatures rise, floods, hurricanes, wildfires, heatwaves, and droughts are becoming more frequent and severe, directly influencing how PFAS move through the environment and accumulate in ecosystems. The increasing volatility of weather patterns has intensified the mobilization of PFAS from contaminated sites, overwhelming existing water treatment infrastructure, and complicating long-term remediation efforts.

Flooding is one of the most significant and immediate climate-driven threats that worsens PFAS contamination. When floodwater inundates landfills, industrial sites, military bases, and wastewater treatment plants, they leach large quantities of PFAS into surface water and groundwater sources (Evich et al., 2022; Schwichtenberg et al., 2023). This contamination is particularly concerning in areas with historical PFAS use, such as fire training sites, airports, and chemical manufacturing facilities, where high concentrations of PFAS are already present in the soil. As climate change increases the frequency of extreme rainfall and storm surges, these flood events are likely to become a major driver of PFAS pollution, introducing contaminants into previously unaffected water systems and making cleanup efforts even more challenging.

Coastal cities and low-lying communities face additional threats, such as rising sea levels and hurricanes push PFAS-laden floodwaters into municipal water supplies. Many coastal areas already struggle with saltwater intrusion into freshwater resources, and PFAS contamination further complicates water management efforts. Storm surges and high tides can carry contaminated sediments from industrial zones into drinking water reservoirs, increasing exposure risks for millions of people (Wee & Aris, 2023). In some cases, the destruction of waste storage sites and aging chemical infrastructure during storms has led to large-scale releases of hazardous PFAS, with effects persisting long after the floodwaters recede.

Wildfires also contribute to PFAS pollution in unexpected and highly concerning ways. Firefighting foams containing PFAS have been widely used in wildfire suppression efforts, particularly in fire-prone regions like California, Australia, and the Mediterranean. As climate change fuels more intense and frequent wildfires, the reliance on PFAS-based fire retardants is likely to increase, leading to greater environmental contamination (Blum et al., 2021). These foams, designed to suppress fires quickly and effectively, can seep into soil and water systems, leaving behind persistent contamination that remains long after the fire is extinguished.

Beyond the direct use of firefighting foams, wildfires themselves can alter PFAS contamination pathways. Research suggests that high temperatures from wildfires can break down PFAS-containing materials, releasing toxic PFAS gases into the atmosphere (Dale et al., 2022). These gases can travel long distances before settling back onto land and water surfaces, contributing to widespread contamination in regions far from the original fire. In addition, burned soil loses its natural ability to filter contaminants, meaning that rainfall following a wildfire can wash PFAS and other toxic chemicals directly into nearby rivers, lakes, and groundwater sources. This cascading effect underscores how climate-driven wildfire activity is accelerating the spread of PFAS pollution in ways that were previously underestimated.

Heatwaves and prolonged drought conditions further complicate PFAS contamination by increasing chemical concentrations in shrinking water sources and reducing the effectiveness of treatment processes. As reservoirs and lakes shrink due to excessive evaporation, the concentration of PFAS in these water bodies increases, making human exposure risks even greater (Wee & Aris, 2023). This problem is particularly severe in regions already struggling with water scarcity, where communities are forced to rely on highly contaminated water sources for drinking, irrigation, and industrial use.

Additionally, heat stress on wastewater treatment plants can reduce the efficiency of PFAS removal technologies, further increasing the risk of contaminated water being discharged back into the environment. Many existing treatment facilities were not designed to handle PFAS contamination, and rising temperatures may further impair their ability to remove these chemicals effectively. In areas where wastewater is recycled for drinking water purposes, this presents a major public health risk, as higher concentrations of PFAS may end up in treated water supplies.

As climate change continues to alter environmental conditions worldwide, the interactions between extreme weather events and PFAS pollution will become more pronounced and difficult to control. The intensification of floods, hurricanes, wildfires, and heatwaves will accelerate the spread of PFAS, making containment and cleanup increasingly difficult. Addressing these challenges requires a multifaceted approach, including more resilient water infrastructure, stricter regulations on PFAS use, improved treatment technologies, and proactive climate adaptation strategies. Without immediate and coordinated action, climate-driven PFAS contamination will continue to threaten public health, ecosystems, and global water security for decades to come.

Challenge of Protecting Water Resources in a Warming World

As climate change intensifies disruptions to global water systems, the challenge of removing PFAS from drinking water supplies becomes even more urgent and complex. Rising temperatures, shifting precipitation patterns, and extreme weather events are placing additional strain on already vulnerable water resources, making it increasingly difficult for existing treatment systems to effectively filter out PFAS contaminants. While technologies such as activated carbon filtration, ion exchange resins, and reverse osmosis have been shown to remove PFAS from drinking water, these methods have become more expensive, less efficient, and harder to sustain as climate change driven events increase contamination levels and put greater pressure on municipal water supplies (Sims et al., 2025).

Droughts, which are becoming longer and more intense worldwide, pose a significant challenge for PFAS management. In drought-stricken regions, shrinking freshwater reserves force communities to rely on alternative water sources, such as groundwater wells, desalination plants, and reclaimed water systems, many of which have higher PFAS contamination levels due to historical pollution and industrial runoff (Evich et al., 2022; Schwichtenberg et al., 2023). The depletion of natural freshwater supplies concentrates PFAS in remaining water sources, increasing human exposure risks and making contamination more difficult to dilute according to Sims et al. (2025). Moreover, their research into low water levels in reservoirs and lakes have shown higher PFAS-laden sediments are disturbed, leading to resuspend trapped PFAS compounds back into the water column, further exacerbating contamination concerns.

Another major concern is the impact of rising sea levels on PFAS contamination, particularly in coastal communities that rely on groundwater aquifers for drinking water. As ocean levels rise, saltwater intrusion into coastal groundwater supplies is becoming a growing threat, altering water chemistry and making conventional treatment processes less effective. Studies suggest that PFAS interact with saltwater differently than freshwater, potentially increasing their bioavailability and toxicity (Cousins et al., 2022). This means that PFAS compounds may behave unpredictably in saline conditions, making it more difficult to design effective water treatment strategies for coastal regions. Moreover, storm surges and coastal flooding events can push PFAS-laden waters into municipal supply systems, overwhelming filtration infrastructure and further complicating cleanup efforts (Sims et al., 2025).

The increased frequency of extreme rainfall events and flooding further threatens water quality by mobilizing PFAS-contaminated sediments into surface water supplies. Floodwaters have been known to wash PFAS from contaminated industrial zones, landfills, and military sites into rivers, lakes, and drinking water reservoirs, rapidly spreading contamination across large geographic areas. In some cases, overwhelmed water treatment facilities may discharge untreated or inadequately treated water back into the environment, allowing PFAS to re-enter drinking water supplies (Schwichtenberg et al., 2023).

To address these growing challenges, governments and water management agencies must develop climate-resilient strategies for PFAS removal. This includes investing in advanced filtration technologies, such as high-capacity activated carbon systems and novel ion exchange resins that can function efficiently under varying environmental conditions. Stricter regulations on PFAS emissions and industrial discharges must also be enforced to prevent additional contamination from entering water systems. Additionally, PFAS monitoring must be integrated into broader climate adaptation plans, ensuring that municipal water supplies remain safe even in the face of climate-driven disruptions (Sims et al., 2025).

Without proactive and large-scale intervention, PFAS contamination will continue to escalate alongside climate change, posing a severe and long-term threat to global water security. The combination of increasing contamination, more extreme weather events, and strained water infrastructure will make it even harder to protect drinking water supplies, ecosystems, and public health. Addressing this crisis requires immediate action, significant investment in sustainable water treatment solutions, and stronger regulatory enforcement to ensure that future generations are not left to grapple with an even more toxic and unstable environment.

Race Against Time

The intersection of PFAS contamination and climate change is more than an environmental issue, it is a defining crisis of our time. As climate change intensifies extreme weather events, disrupts water systems, and accelerates chemical movement, it exacerbates the dangers of PFAS pollution, turning an already widespread contamination problem into a global catastrophe. With floods dispersing toxic sediments, wildfires releasing PFAS-laden smoke into the air, and droughts concentrating contamination in shrinking water supplies, the window for effective intervention is closing rapidly.

This crisis demands more than piecemeal regulatory efforts or reactionary cleanup strategies. It requires bold, decisive action—governments must enforce stricter PFAS regulations, water management systems must be re-engineered for resilience, and scientific research must push the boundaries of PFAS remediation technologies. More importantly, PFAS mitigation must be integrated into climate change adaptation policies, ensuring that communities already on the frontlines of rising seas, extreme storms, and water scarcity are not further burdened by toxic exposure.

The failure to act swiftly will leave future generations with a poisoned planet—where every drop of water, every breath of air, and every bite of food carries the lasting legacy of unchecked pollution. The challenge before us is clear: either we confront PFAS contamination with the urgency it demands, or we allow it to become another irreversible environmental tragedy in the wake of climate change. The time for half-measures is over. The time for action is now.

Chapter 17

Innovations in Mitigation and Green Chemistry

The growing recognition of PFAS contamination as a global environmental and public health crisis has driven urgent research into innovative mitigation strategies and sustainable alternatives. With traditional remediation methods often proving costly, inefficient, or limited in scope, scientists and engineers are developing new technologies to remove PFAS from contaminated water, soil, and air. At the same time, green chemistry is revolutionizing industrial practices, leading to the creation of PFAS-free alternatives for consumer and commercial applications.

As industries, governments, and consumers push for stricter PFAS regulations, the demand for certified PFAS-free products is rising. This shift signals a major transition toward sustainability, where companies must demonstrate compliance with environmental standards and commit to safer, non-toxic alternatives. This chapter explores emerging PFAS removal technologies, advancements in green chemistry, and the future of PFAS-free certifications that could reshape how industries and consumers approach chemical safety.

Emerging Technologies for PFAS Removal

Traditional methods for PFAS removal, such as granular activated carbon (GAC) filtration, ion exchange resins, and reverse osmosis, have been widely used to treat contaminated water sources (Wee & Aris, 2023). While effective, these approaches are energy-intensive, costly, and often struggle to handle long-chain and emerging short-chain PFAS compounds. As a result, researchers have focused on developing novel and more sustainable PFAS removal technologies.

One of the most promising innovations in PFAS remediation is bioremediation, which utilizes microorganisms or enzymes to break down PFAS compounds into non-toxic byproducts. Some bacteria and fungi have demonstrated the ability to degrade specific PFAS molecules, offering a potential low-cost and environmentally friendly solution (Sims et al., 2025). While bioremediation remains in early stages of research, laboratory studies have identified engineered microbes capable of accelerating PFAS degradation under controlled conditions.

Another breakthrough technology involves electrochemical and plasma-based destruction methods. Electrochemical oxidation applies high-voltage currents to break apart PFAS molecules, reducing them into harmless carbon and fluorine ions (Evich et al., 2022; Schwichtenberg et al., 2023). Similarly, plasma reactors generate high-energy plasma fields that disrupt PFAS molecular structures, effectively neutralizing them. These technologies show promise for on-site remediation, particularly in highly contaminated areas like industrial discharge zones and firefighting training sites.

Nanotechnology is also playing a critical role in PFAS removal. Advanced nanomaterials, such as carbon nanotubes, metal-organic frameworks (MOFs), and molecularly imprinted polymers, have demonstrated high efficiency in capturing and degrading PFAS molecules in water (Cousins et al., 2022). These materials work by adsorbing PFAS with extreme precision, preventing them from reentering the environment. While still in experimental stages, nanotechnology-driven filtration systems could provide a scalable and cost-effective alternative to conventional treatment methods.

Despite these advancements, PFAS remediation faces several key challenges. The sheer persistence of PFAS means that complete degradation remains difficult, and scaling up these emerging technologies for widespread application requires further investment and regulatory approval. However, continued innovation in removal technologies offers a hopeful path forward in reducing PFAS contamination in air, water, and soil.

Advances in Green Chemistry to Replace PFAS in Industrial Applications

While PFAS removal technologies focus on cleaning up existing contamination, green chemistry aims to prevent pollution at its source by developing safer, non-toxic alternatives. Industries that have traditionally relied on PFAS for waterproofing, stain resistance, and heat stability are now investing in next-generation materials that offer similar performance without environmental persistence (Blum et al., 2021).

One major area of innovation is the development of bio-based and fluorine-free coatings for consumer goods. Researchers have designed plant-derived wax coatings and silica-based repellents that replicate PFAS-like properties without their long-term environmental impact. Companies in the outdoor apparel, food packaging, and cookware industries have already started transitioning to fluorine-free alternatives, demonstrating that green chemistry solutions can meet commercial demands while reducing ecological harm (Cousins et al., 2022).

In the electronics sector, manufacturers are exploring PFAS-free circuit boards and semiconductors, replacing PFAS-based insulating materials with silicone-based and ceramic alternatives. Given the growing demand for sustainable electronics, this shift could have a significant impact on reducing PFAS use in high-tech industries (Dale et al., 2022).

Additionally, the search for PFAS-free firefighting foams has led to the development of protein-based and biodegradable foams that meet fire safety regulations without releasing persistent pollutants into the environment (Johns Hopkins APL, 2023; University of Maine News, 2025). While testing and certification processes remain ongoing, several airport and military facilities have already adopted fluorine-free foams, proving that green chemistry solutions can replace PFAS in high-risk applications.

Despite these advancements, challenges remain in scaling up production, ensuring cost-effectiveness, and overcoming industry resistance. Many businesses still prefer PFAS due to cost efficiency and long-standing supply chains, making policy incentives and consumer demand essential drivers of change. The continued progress of green chemistry initiatives will be crucial in eliminating PFAS reliance across industries and preventing future contamination.

Future of PFAS-Free Certifications and Sustainable Alternatives

As awareness of PFAS risks grows, PFAS-free certifications are becoming an important tool for consumers, manufacturers, and regulators. Certifications help distinguish products that do not contain PFAS, encouraging businesses to transition toward safer alternatives. Organizations such as GreenScreen Certified, the Environmental Working Group (EWG), and the EPA's Safer Choice program have established PFAS-free labeling criteria, allowing consumers to make informed purchasing decisions (Evich et al., 2022; Schwichtenberg et al., 2023).

In response to growing regulatory pressures, some industries have voluntarily phased out PFAS, opting for certified PFAS-free materials in textiles, food packaging, and cosmetics. However, greenwashing remains a significant concern, as some companies falsely claim to be PFAS-free while still using chemically similar compounds with untested toxicity (Blum et al., 2021). Stronger certification standards, third-party testing, and supply chain transparency are needed to ensure true PFAS elimination from consumer products.

The future of sustainable PFAS alternatives depends on continued investment in research, regulatory support, and consumer awareness. As more businesses commit to PFAS-free production, the market for safer, environmentally responsible alternatives will expand, further driving innovation in green chemistry. If PFAS-free certifications become industry norms, they could push companies worldwide to phase out these harmful chemicals permanently.

A Defining Moment for a PFAS-Free Future

The fight against PFAS contamination has reached a critical turning point. No longer a problem confined to scientific circles or regulatory agencies, the urgency of eliminating PFAS has entered mainstream discourse, with industries, governments, and consumers recognizing the devastating consequences of inaction. The good news is that breakthroughs in mitigation technologies and green chemistry innovations are proving that a PFAS-free future is within reach. Bioremediation, electrochemical degradation, and nanotechnology are revolutionizing how we remove PFAS from contaminated environments, while green chemistry is leading the way in creating safer, high-performance alternatives that no longer rely on these toxic compounds.

But progress is not guaranteed. For these solutions to make a lasting impact, regulatory enforcement must be strengthened, financial incentives must accelerate innovation, and corporate responsibility must become the norm, not the exception. Consumers must demand accountability, refusing to support industries that continue to prioritize profit over public health. The transition away from PFAS is no longer a question of feasibility, it is an absolute necessity.

This is a defining moment in chemical safety and environmental responsibility. The choices we make today will shape the health of future generations and determine whether we break free from a legacy of persistent pollution or allow it to remain an enduring threat. The time for small, incremental steps has passed. We must act decisively, innovate relentlessly, and commit fully to eliminating PFAS before they cause irreparable harm. The future is watching—will we rise to the challenge?

.

Chapter 18

Ethical and Cultural Implications

With the widespread contamination of per- and polyfluoroalkyl substances (PFAS) raises ethical and cultural concerns that extend beyond environmental science and regulatory policies. PFAS pollution disproportionately affects marginalized communities, Indigenous populations, and economically disadvantaged regions, raising serious questions about environmental justice, equity in cleanup efforts, and the balance between economic development and environmental protection.

Despite overwhelming evidence linking PFAS exposure to severe health risks, including cancer, endocrine disruption, and immune system suppression, affected communities often lack the resources, political influence, and legal power to demand accountability from polluting industries (Grandjean & Clapp, 2015). This disparity highlights ethical concerns about who bears the brunt of environmental contamination and who benefits from industrial and economic development. Furthermore, the slow pace of PFAS cleanup efforts and unequal access to safe drinking water underscore systemic inequities in environmental health protection.

As governments, corporations, and advocacy groups work to mitigate PFAS contamination, they must prioritize environmental justice, ensure equitable access to clean water, and create policies that balance economic progress with long-term sustainability. Failing to do so risks deepening existing social and environmental inequalities, leaving vulnerable populations to suffer the consequences of unchecked pollution.

Impacts of PFAS on Marginalized and Indigenous Communities

Environmental justice is a fundamental human rights issue, ensuring that all communities, regardless of income, race, or location, have equal access to a clean and healthy environment. However, the reality is that low-income communities, Indigenous populations, and people of color are disproportionately affected by PFAS contamination (Evich et al., 2022; Schwichtenberg et al., 2023). Many of these communities are located near industrial sites, military bases, landfills, and wastewater treatment plants, where PFAS exposure is significantly higher.

Indigenous communities in particular face heightened risks due to their reliance on traditional lands and waterways for food, cultural practices, and daily life. Studies have shown that PFAS levels in drinking water on Native American reservations are significantly higher than in non-tribal areas, largely due to historical land use policies that have forced Indigenous communities to live near industrial waste sites or military facilities that used PFAS-containing firefighting foams (Cousins et al., 2022).

In Alaska, where many Indigenous communities depend on subsistence fishing and hunting, PFAS bioaccumulation in fish and wildlife poses a severe health threat. Exposure through traditional food sources not only endangers physical health but also threatens cultural preservation, as these practices are deeply tied to spiritual and community identity (Blum et al., 2021).

In addition to Indigenous populations, low-income and rural communities across the United States and globally suffer from higher-than-average PFAS contamination levels. These communities often lack financial and political power to fight back against polluting industries, leaving them at greater risk of exposure. Industrial zones in states like Michigan, West Virginia, and North Carolina have recorded some of the highest PFAS concentrations in drinking water, disproportionately affecting communities of color and economically disadvantaged households (Toxics Use Reduction Institute (TURI), 2025).

The ethical question remains: Why do some communities continue to suffer from toxic exposure while wealthier areas see quicker cleanup responses? The answer lies in historical patterns of environmental racism, economic inequity, and corporate negligence, all of which must be addressed to create meaningful, long-term solutions to PFAS contamination.

Equity in Cleanup Efforts and Access to Safe Water

Ensuring equity in PFAS cleanup efforts and universal access to safe drinking water is one of the greatest challenges in environmental policy and public health. Although major corporations like 3M and DuPont have settled lawsuits related to PFAS pollution, the distribution of remediation funds has not been equal, and many contaminated communities still lack access to clean water or resources for long-term solutions (Wee & Aris, 2023).

One of the biggest obstacles to equitable cleanup efforts is the cost of PFAS remediation technologies. Advanced filtration systems, such as granular activated carbon (GAC), ion exchange resins, and reverse osmosis, are effective but extremely expensive (Toxics Use Reduction Institute (TURI), 2025). Many low-income communities cannot afford these technologies, leaving them dependent on outdated or insufficient filtration methods. This disparity means that while wealthier municipalities can invest in state-of-the-art PFAS removal infrastructure, economically disadvantaged areas remain at risk of prolonged exposure.

Government-led cleanup programs have also demonstrated geographical and political bias. Some contaminated regions receive immediate state or federal funding for water treatment and soil remediation, while others face delays due to bureaucratic inefficiencies, lack of media attention, or political gridlock (Schwichtenberg et al., 2023). This disparity raises an ethical concern: Should access to clean water depend on political influence, financial resources, or legal action? If public health is truly a fundamental right, then every community should have equal access to remediation efforts, regardless of income or location.

Internationally, the PFAS crisis mirrors broader global water security challenges. In developing nations, where industrial pollution regulations are often weaker, PFAS contamination has been largely unaddressed, leaving millions exposed to toxic chemicals in drinking water. Countries in Southeast Asia, South America, and Africa, where multinational corporations have offshored industrial production, now face some of the highest PFAS contamination levels worldwide with limited resources to address the crisis (Cousins et al., 2022).

Addressing these inequities in PFAS cleanup requires a coordinated, global approach. Governments must prioritize funding for at-risk communities, international organizations must support sustainable water treatment solutions, and corporations responsible for PFAS pollution must be held accountable for remediation costs.

Balancing Economic Development with Environmental Protection

One of the most contentious ethical dilemmas surrounding PFAS regulation is the balance between economic development and environmental protection. PFAS-based chemicals have been integral to numerous industries, including aerospace, medical technology, construction, and consumer goods (Blum et al., 2021). While the transition to PFAS-free alternatives is possible, many companies argue that immediate bans or strict regulations could hinder economic growth, drive up costs, and eliminate jobs.

Corporations often frame PFAS regulation as a trade-off between public health and economic prosperity, but history has demonstrated that short-term economic gains at the expense of environmental responsibility led to long-term financial and health crises. The cost of cleaning up PFAS contamination, addressing public health impacts, and compensating affected communities far outweighs the cost of transitioning to safer alternatives (Schwichtenberg, et al. 2023; Toxics Use Reduction Institute (TURI), 2025).

Many argue that the real economic concern is not the loss of PFAS-dependent industries but the potential financial collapse of communities suffering from contamination. When local water supplies are polluted beyond safe limits, property values decline, businesses relocate, and public health costs skyrocket. Investing in sustainable alternatives and responsible corporate practices is not just an ethical obligation, it is an economic necessity for long-term stability.

Governments must enforce stricter regulations that prioritize human health without stifling economic innovation. Financial incentives, tax breaks, and grants can help businesses transition away from PFAS while ensuring that workers are not displaced.

A Call for Justice and Responsibility

The ethical and cultural implications of PFAS contamination demand urgent action and unwavering commitment to justice. This is not just an environmental crisis, it is a human rights issue. Communities that have historically borne the brunt of industrial pollution can no longer be ignored. Access to clean water should never be a privilege dictated by wealth or political influence, it is a fundamental right that must be protected for all people, regardless of race, income, or geography.

Balancing economic growth with environmental responsibility is not an impossible task, it is an ethical imperative. Industries must acknowledge their role in this crisis and invest in sustainable alternatives, rather than prioritizing short-term profits over public health and ecological integrity. Governments must strengthen regulations, enforce corporate accountability, and ensure that the most vulnerable communities are no longer sacrificed for industrial expansion.

The fight against PFAS is about more than science and policy is about fairness, responsibility, and the moral obligation to protect future generations. The choices made today will define the legacy of environmental ethics in the decades to come. Will society continue to turn a blind eye to those suffering the most, or will it finally uphold the principles of justice, equity, and sustainability? The answer will shape not only our present actions but the world we leave behind. The time to act is now.

Chapter 19

Lessons from Global Successes

As the United States grapples with the growing crisis of PFAS contamination, other nations have already taken decisive action to regulate, remediate, and prevent the spread of these persistent pollutants. Countries such as Denmark, Sweden, Australia, and the Netherlands have implemented aggressive PFAS bans, comprehensive cleanup efforts, and industry-wide reforms that serve as models for the rest of the world. These international successes provide critical lessons for U.S. policymakers, industry leaders, and environmental advocates striving to combat PFAS pollution.

Beyond individual national efforts, international collaboration has played a crucial role in addressing PFAS contamination. Agreements such as the Stockholm Convention on Persistent Organic Pollutants (POPs) have helped establish global frameworks for regulating hazardous chemicals, fostering cooperation among nations to phase out the most dangerous PFAS compounds (Cousins et al., 2022). By examining effective PFAS regulations, cleanup strategies, and cross-border initiatives, the U.S. can adopt proven policies and accelerate progress toward a PFAS-free future.

There are many case studies of effective PFAS regulation and cleanup efforts from across the globe, here are four instances:

Denmark: The First Nation to Ban PFAS in Food Packaging

Denmark has taken one of the strongest stances against PFAS contamination, becoming the first country in the world to implement a nationwide ban on PFAS in food packaging (Blum et al., 2021). Recognizing the risk of PFAS leaching from grease-resistant packaging into food, Danish regulators banned all PFAS-based food contact materials in 2020, setting a precedent for other European nations.

The Danish model demonstrates that bold regulatory action is possible when public health is prioritized over corporate interests. The country worked closely with scientists and industry stakeholders to identify viable, PFAS-free alternatives, ensuring that the transition away from PFAS did not disrupt the food industry. This approach has since influenced other EU nations to implement similar food packaging restrictions, highlighting the effectiveness of swift, science-based policymaking (Cousins et al., 2022).

Sweden: Holding Polluters Accountable Through Legal Action

Sweden has taken a litigation-driven approach to PFAS accountability, setting a global precedent for corporate responsibility. In 2021, Swedish authorities filed landmark lawsuits against chemical manufacturers, including 3M and DuPont, seeking compensation for the costs of PFAS cleanup and environmental damage (Evich et al., 2022; Schwichtenberg et al., 2023). The Swedish government has also invested heavily in advanced water treatment technologies, ensuring that municipal water systems are free from PFAS contamination.

This legal strategy serves as a powerful example of how governments can force corporations to take financial responsibility for PFAS pollution. Unlike the U.S. model, which has largely relied on voluntary industry action, Sweden's approach demonstrates that strict enforcement and financial penalties can drive meaningful change. By adopting similar polluter-pays policies, the U.S. could ensure that the costs of PFAS cleanup do not fall on taxpayers.

Australia: Comprehensive Monitoring and Remediation

Australia has implemented one of the world's most comprehensive PFAS monitoring and remediation programs, focusing on military sites, firefighting training facilities, and industrial zones (Wee & Aris, 2023). The Australian government established national PFAS task forces to oversee testing, public health studies, and environmental remediation, ensuring that affected communities receive timely assistance.

A key lesson from Australia's approach is the importance of early detection and rapid response. By prioritizing large-scale PFAS monitoring, the country has been able to contain contamination before it spreads further into drinking water supplies. The U.S. could benefit from expanding its PFAS surveillance programs to ensure early intervention and prevent long-term exposure risks.

The Netherlands: Leading Green Chemistry and PFAS Alternatives

The Netherlands has emerged as a global leader in green chemistry, actively promoting the development of PFAS-free industrial processes. Dutch scientists and businesses have collaborated to create fluorine-free coatings, textiles, and fire suppression foams, demonstrating that PFAS are not essential for high-performance materials.

The Dutch model highlights the importance of industry innovation in solving the PFAS crisis. By offering financial incentives for companies to develop sustainable alternatives, the Netherlands has accelerated the transition away from PFAS without harming economic growth. The U.S. could follow this example by investing in PFAS-free research and development, providing grants and tax incentives to companies that commit to eliminating PFAS from their supply chains.

What the U.S. Can Learn from Other Countries

The United States has lagged behind many developed nations in PFAS regulation, largely due to industry lobbying, weak federal mandates, and inconsistent enforcement (Blum et al., 2021). However, the successes of Denmark, Sweden, Australia, and the Netherlands offer valuable lessons for improving U.S. policies.

One of the most critical takeaways is that strong, enforceable bans, like Denmark's PFAS-free food packaging law, are both practical and effective. If the U.S. were to implement a nationwide ban on non-essential PFAS use, it could prevent future contamination and drive industries toward safer alternatives.

The Swedish model demonstrates the power of litigation in holding corporations accountable. If the U.S. expanded its legal efforts against PFAS manufacturers, it could recover billions in cleanup costs and deter future chemical pollution.

Australia's focus on nationwide PFAS monitoring underscores the importance of proactive surveillance. If the U.S. expanded PFAS testing in drinking water supplies, soil, and air, it could identify contamination hotspots earlier and reduce long-term exposure risks.

The Netherlands provides a blueprint for transitioning to PFAS-free alternatives. By investing in green chemistry and rewarding companies that develop sustainable materials, the U.S. could accelerate the shift away from PFAS-dependent industries.

International Collaboration in Tackling PFAS Contamination

Because PFAS are global pollutants that travel through air, water, and food chains, international cooperation is critical to solving the crisis.

Several global agreements and research partnerships have helped coordinate PFAS reduction efforts across borders.

The Stockholm Convention on Persistent Organic Pollutants (POPs) has been instrumental in phasing out some of the most hazardous PFAS chemicals (Cousins et al., 2022). While the U.S. has been reluctant to fully commit to the treaty, increased participation could strengthen global PFAS regulations and drive more aggressive cleanup efforts.

Research collaborations between the European Union, Australia, and the U.S. have facilitated breakthroughs in PFAS detection and remediation technologies. Expanding international data-sharing and joint research projects could help the U.S. access cutting-edge solutions faster.

Additionally, cross-border legal actions against PFAS manufacturers—such as Sweden's lawsuits against 3M—could set global legal precedents that force chemical companies to clean up contamination worldwide (Schwichtenberg et al., 2023).

By actively engaging in international PFAS policy discussions, the U.S. could play a leadership role in banning PFAS production and improving global environmental protections.

A Defining Moment for Global Leadership

The fight against PFAS contamination is not just an environmental challenge, it is a test of global resolve, scientific ingenuity, and moral responsibility. Nations like Denmark, Sweden, Australia, and the Netherlands have proven that decisive action is possible. They have implemented strict regulations, held corporations accountable, and invested in scientific innovation to protect their citizens and ecosystems. Their successes are not anomalies, they are blueprints for what must be done on a larger scale.

The United States now stands at a crossroads. Will it continue to allow industrial influence and bureaucratic inertia to delay progress, or will it learn from these global successes and take a leadership role in eliminating PFAS contamination? The answer will determine not only the health of its people and environment but also its credibility as a nation committed to sustainability and public welfare.

This is a defining moment—a chance to transform policy, enforce corporate responsibility, and invest in safer alternatives. The cost of inaction is immeasurable, both in human lives and environmental damage. The U.S. has the opportunity, and the obligation, to act with urgency, integrity, and ambition. The world is watching. The time to act is not tomorrow. It is now.

Chapter 20

The Path Forward

The battle against PFAS contamination is at a turning point. Years of scientific research, legal actions, and public advocacy have exposed the true scale of this crisis, yet meaningful regulatory action remains slow and fragmented. While some progress has been made, the persistent and bioaccumulative nature of PFAS demands more than incremental policy shifts—it requires a bold, science-driven transformation in how we regulate chemicals, protect public health, and prevent future environmental disasters.

The path forward must be rooted in evidence-based policies that prioritize human and environmental safety over corporate interests. Achieving this goal requires bridging the gap between research, regulation, and public action, ensuring that scientific discoveries translate into enforceable laws and community-driven solutions. A truly sustainable future not only eliminates PFAS contamination but also establishes safeguards to prevent similar chemical crises from emerging in the future.

Recommendations for Science-Based Policy Changes

For decades, chemical regulations have lagged behind scientific discoveries, allowing industries to continue producing and using harmful substances long after their dangers were well-documented. The failure of existing policies to prevent widespread PFAS contamination underscores the urgent need for science-based reforms that reflect the latest research on toxic exposure, environmental persistence, and human health impacts (Grandjean & Clapp, 2015).

One of the most critical steps in PFAS regulation is the adoption of a class-based approach, treating all PFAS chemicals as a single group rather than regulating them one by one (Cousins et al., 2022). Historically, chemical manufacturers have evaded regulation by slightly altering PFAS structures, creating new compounds that remain just as persistent and toxic but fall outside the scope of existing bans. A class-based regulatory framework would eliminate this loophole, ensuring that all PFAS compounds, both legacy and emerging, are restricted unless proven safe.

Additionally, PFAS policy must shift from reactive to preventive regulation. Many regulatory agencies, including the U.S. Environmental Protection Agency (EPA), have historically relied on post-market evaluations, meaning that chemicals are introduced into commercial use before their long-term health effects are fully understood (Grandjean & Clapp, 2015). This approach has allowed PFAS and countless other hazardous substances to enter the environment unchecked, forcing governments to clean up contamination after the damage has already been done. A precautionary regulatory framework, similar to the European Union's REACH (Registration, Evaluation, Authorization and Restriction of Chemicals) program, would require chemical manufacturers to prove the safety of their products before they are allowed on the market.

Another essential policy shift is the strict enforcement of corporate accountability laws. While some PFAS manufacturers, such as 3M and DuPont, have faced lawsuits and settlements, many continue to deny full responsibility for decades of pollution (Schwichtenberg et al., 2023). Governments must expand polluter-pays policies, ensuring that companies responsible for contamination bear the financial burden of cleanup, rather than taxpayers. Legislation should eliminate legal loopholes that allow corporations to settle lawsuits without admitting fault, preventing future chemical disasters from unfolding under the same pattern of industry deception.

Finally, scientific funding for PFAS research and remediation technologies must be significantly expanded. While promising PFAS removal methods, such as bioremediation and electrochemical degradation, are in development, most remain too costly or experimental for large-scale deployment (Wee & Aris, 2023). Government agencies must invest heavily in advancing and scaling up these technologies, ensuring that contaminated communities have access to effective, affordable cleanup solutions. Without substantial public investment, the burden of innovation will remain on private research institutions and environmental organizations, slowing progress toward long-term remediation goals.

Bridging the Gap Between Research, Regulation, and Public Action

The failure to regulate PFAS effectively is not due to a lack of scientific knowledge, but rather a failure to translate research into enforceable policy. This gap between scientific discovery, governmental regulation, and public awareness has allowed PFAS contamination to spread unchecked for decades, even as evidence of its dangers became impossible to ignore (Cousins et al., 2022). Closing this gap requires better communication, increased public engagement, and stronger collaborations between scientists, policymakers, and advocacy groups.

One major obstacle to effective PFAS regulation is corporate influence over regulatory agencies. Chemical industry lobbying has delayed PFAS bans, weakened proposed legislation, and funded misleading studies that cast doubt on scientific findings (Schwichtenberg et al., 2023). To counteract this influence, regulatory agencies must operate with greater independence and transparency, ensuring that scientific integrity is prioritized over industry interests. Governments should implement conflict-of-interest policies, preventing chemical manufacturers from funding the very research used to regulate their products.

Additionally, public pressure remains one of the most effective tools for driving regulatory action. Many of the strongest PFAS bans and cleanup efforts—such as those in Denmark, Sweden, and several U.S. states—were implemented in response to public outcry and grassroots advocacy (Blum et al., 2021). Educating communities about PFAS risks and mobilizing public demand for policy change is essential for ensuring that governments act with urgency. Public engagement campaigns, environmental justice movements, and media coverage must continue to push PFAS reform into the political spotlight, making it impossible for lawmakers to ignore.

A stronger bridge between research and public action also requires clearer risk communication from scientists and policymakers. Many people remain unaware of PFAS contamination in their drinking water, consumer products, and local environments, largely because government agencies have failed to provide accessible, transparent information (Blum et al., 2021). Scientists must work with journalists, educators, and advocacy groups to ensure that PFAS-related health risks and policy developments are communicated effectively to the public.

Building a Sustainable Future and Preventing Future Chemical Crises

The PFAS crisis is not just a failure of chemical regulation, it is a symptom of a broken system that allows hazardous substances to be mass-produced without adequate oversight. Preventing future chemical crises requires a fundamental shift in how we evaluate, regulate, and replace hazardous materials before they cause irreversible environmental and health damage.

One of the most promising solutions is the advancement of green chemistry—an approach that prioritizes the development of safer, biodegradable, and non-toxic materials to replace hazardous chemicals (Blum et al., 2021). Governments must fund and incentivize the development of PFAS-free alternatives, ensuring that industries have financial and regulatory motivations to transition toward safer materials. Investment in green chemistry education is also critical, ensuring that future generations of scientists and engineers are trained to design chemicals with sustainability in mind.

Additionally, long-term chemical oversight mechanisms must be restructured. A key lesson from the PFAS crisis is that waiting until contamination reaches catastrophic levels before taking action is both costly and deadly. Regulatory agencies should adopt a proactive approach to chemical safety, implementing routine environmental monitoring, stronger testing requirements, and stricter chemical approval processes (Cousins et al., 2022).

Finally, global cooperation is essential. PFAS contamination is a transboundary issue, affecting water, air, and food systems beyond national borders. International agreements, such as the Stockholm Convention on Persistent Organic Pollutants, should be expanded to include stricter PFAS restrictions, ensuring that no country serves as a dumping ground for toxic chemicals (Schwichtenberg et al., 2023).

A Defining Moment for Chemical Safety and Public Health

The PFAS crisis is more than an environmental catastrophe, it is a glaring indictment of systemic failures in chemical regulation, corporate ethics, and public health protections. It has laid bare the consequences of unchecked industrial influence, weak regulatory oversight, and the prioritization of profit over human well-being. If history has taught us anything, it is that inaction has a cost—one measured in contaminated water, poisoned ecosystems, and lives cut short by preventable diseases.

This moment demands more than half-measures and delayed reforms. Policymakers must adopt science-based regulations, hold corporations accountable for the damage they have caused, and accelerate investment in safer alternatives. The responsibility to bridge the gap between research, policy, and public action is not just a task for scientists or legislators, it is a collective duty for society as a whole.

The legacy of PFAS contamination must serve as an urgent warning. If we fail to act decisively now, we will repeat the same mistakes with the next generation of hazardous chemicals, condemning future generations to the same cycles of pollution, litigation, and irreversible harm. But the path forward is clear. We know the dangers, we know the solutions, and we know what must be done.

Chapter 21

Reflections on the Scale of
Our PFAS Legacy

The PFAS crisis is not just a story of scientific discovery, corporate negligence, or regulatory failure—it is a defining moment in the way we choose to govern chemicals, protect our environment, and safeguard public health. These "forever chemicals" have infiltrated nearly every corner of our planet—from the drinking water of small-town communities to the blood of newborns, from Arctic ice sheets to deep ocean currents. What began as an industrial marvel has now become one of the most profound environmental and public health disasters of our time.

Yet, this is not where the story has to end. The lessons of PFAS are not just about contamination; they are about transformation. This crisis has given us a rare opportunity to confront systemic failures and demand something better. It has shown us the cost of complacency, the dangers of industry-driven science, and the catastrophic consequences of placing economic convenience over environmental safety. But it has also revealed the power of science, the resilience of communities, and the ability of informed citizens to push for real change.

A Call to Action for Policymakers, Scientists, Industries, and Individuals

The PFAS crisis is not an inevitable fate, it is a challenge we have the power to solve. The path forward is not inaction but bold and urgent steps to undo the damage and prevent future generations from inheriting an even greater burden.

For policymakers, this is the moment to rewrite the rules—to move beyond weak, piecemeal regulations and implement science-based, enforceable laws that hold polluters accountable and eliminate PFAS at the source. The patchwork of state-led bans and industry loopholes must end. PFAS must be regulated as a class, not one-by-one, and we must shift from reacting to crises to preventing them altogether.

For scientists and researchers, the challenge is clear: continue pushing the boundaries of discovery. We need breakthroughs in PFAS remediation, advancements in green chemistry, and stronger interdisciplinary collaboration to not only clean up this mess but also ensure it is never repeated. The role of science must not be to justify past damage but to forge a future where safer, sustainable alternatives are the standard.

For industries, the era of chemical impunity is over. Companies that continue to manufacture, use, and hide PFAS contamination will face increasing scrutiny, legal battles, and public backlash. The industries that embrace change, invest in sustainable alternatives, and commit to corporate accountability will lead the future. The question for every manufacturer today is: Do you want to be part of the solution or forever linked to one of the greatest environmental scandals in history?

For individuals, the power to change the course of this crisis lies in awareness, activism, and daily choices. Consumers must demand transparency from corporations, push for PFAS-free products, and support policies that prioritize health over profit. The pressure from grassroots movements, community organizations, and environmental advocates has already driven some of the strongest PFAS regulations to date. This is proof that public outrage, when organized and persistent, can force even the most powerful industries to change.

A Hidden Toxic Threat: The Case of Perfluorodecanoic Acid (PFDA)

Among the many PFAS compounds wreaking havoc on human health, perfluorodecanoic acid (PFDA) is particularly concerning. PFDA, a long-chain perfluorinated carboxylic acid, has been shown to have significant toxicological effects, including severe endocrine disruption, immune suppression, and reproductive toxicity (Nguyen et al., 2020). This compound, like others in the PFAS family, binds strongly to serum proteins and accumulates in vital organs, particularly the liver and kidneys, where it can cause long-term damage (Domingo et al., 2021).

Studies have demonstrated that PFDA exposure disrupts thyroid hormone regulation, which is essential for metabolic health, brain function, and fetal development (Grandjean & Clapp, 2015). The chemical's persistence in biological systems has raised alarms about its role in metabolic disorders such as obesity and insulin resistance, exacerbating global public health challenges (Post et al., 2012).

Beyond individual health impacts, PFDA contamination in surface water and groundwater continues to pose risks to ecosystems and drinking water supplies. Because of its high bioaccumulation potential, PFDA poses a particular threat to pregnant women and children, as it can cross the placental barrier and be transferred through breast milk (Domingo et al., 2021). This raises profound concerns about intergenerational toxic exposure, further emphasizing the need for immediate regulatory action and comprehensive research into the long-term effects of PFDA and other PFAS compounds.

Hope for a Cleaner, Safer World

We have crossed the threshold. PFAS, those "forever chemicals" engineered for convenience, are now everywhere. They coat our clothes, line our food packaging, seep into our drinking water, and even dust our roads. We drive on them, eat from them, and breathe them in. The web of PFAS contamination is so vast that no corner of the Earth is untouched. The truth is stark: we are living in a PFAS world, and there is no turning back.

So now, we face a choice. Do we resign ourselves to the consequences of this chemical entanglement, accepting the health risks, environmental devastation, and generational impacts as an unavoidable cost of modern life? Or do we rise to the occasion and commit, on a global scale, to dismantling this toxic legacy?

Though the scale of the PFAS challenge is staggering, hope is not just possible, it is already unfolding. Countries like Denmark and Sweden have banned PFAS in consumer products, major corporations are pledging to phase them out, and scientists are developing breakthrough methods to extract PFAS from water, soil, and air. The next chapter of this story is still being written, and it does not have to be one of further contamination—it can be one of redemption, responsibility, and resilience.

This is our moment to redefine chemical safety, to set a new standard for environmental protection, and to build a future where the mistakes of the past do not dictate the health of the next generation. The PFAS crisis has presented us with a stark choice: continue down the same reckless path, ignoring the warning signs, or forge a new one, where science, justice, and sustainability guide our decisions.

When future generations look back at this moment, will they see a turning point, where humanity finally stood up to unchecked chemical pollution? Or will they see yet another tragedy, another lesson ignored?

The answer is up to us. The time for action is now. The fight for a cleaner, safer world is ours to win.

Finally, PFAS is the single most existential threat to human and environmental health, a relentless, invisible poison seeping into our bodies, our water, and our future. If we do not act now, we are complicit in a legacy of contamination that will outlive us all.

Therefore, the only questions left are:

Do we have the courage to act before it's too late?

or

Is it too late to act and we must now deal with the consequences?

References

3M. (2023). 3M announces $10.3 billion settlement to resolve PFAS water contamination claims. https://www.3m.com

Ahrens, L., & Bundschuh, M. (2014). Fate and effects of poly- and perfluoroalkyl substances in the aquatic environment: A review. Environmental Toxicology and Chemistry, 33(9), 1921–1929. https://doi.org/10.1002/etc.2663

Agency for Toxic Substances and Disease Registry. (2021). Per- and polyfluoroalkyl substances (PFAS) and your health. https://www.atsdr.cdc.gov/pfas/index.html

Agency for Toxic Substances and Disease Registry. (2022). Toxicological profile for perfluoroalkyls. https://www.atsdr.cdc.gov/toxprofiles/tp200.pdf

Agency for Toxic Substances and Disease Registry. (2022). How PFAS impacts your health. https://www.atsdr.cdc.gov/pfas/about/health-effects.html

American Chemical Society. (2023, August 28). The battle over PFAS in Europe. Retrieved from https://cen.acs.org/policy/chemical-regulation/battle-over-PFAS-Europe/101/i31

Andrews, D. Q., & Naidenko, O. V. (2020). Population-wide exposure to per- and polyfluoroalkyl substances from drinking water in the United States. Environmental Science & Technology Letters, 7(12), 931-936. https://doi.org/10.1021/acs.estlett.0c00713

Antea Group. (2022). PFAS regulation around the world. https://int.anteagroup.com/news-and-media/blog/pfas-regulation-around-the-world

Bilott, R. (2019). Exposure: Poisoned water, corporate greed, and one lawyer's twenty-year battle against DuPont. Atria Books.

https://www.amazon.com/Exposure-Poisoned-Corporate-Lawyers-Twenty-Year/dp/1501172816

Blake, M. (2018). The curse of the new Teflon toxins. The Intercept. https://theintercept.com/2018/10/15/teflon-chemicals-genx-water-contamination/

Blum, A., Balan, S. A., Scheringer, M., Trier, X., Goldenman, G., Cousins, I. T., & Diamond, M. L. (2021). The Madrid statement on poly- and perfluoroalkyl substances (PFASs). Environmental Health Perspectives, 123(5), 1-10. https://doi.org/10.1289/ehp.1509934

Butt, C. M., Berger, U., Bossi, R., & Tomy, G. T. (2010). Levels and trends of poly- and perfluorinated compounds in the Arctic environment. Science of the Total Environment, 408(15), 2936–2965. https://doi.org/10.1016/j.scitotenv.2010.03.015

California Department of Toxic Substances Control. (2021, October 5). Food packaging containing perfluoroalkyl or polyfluoroalkyl substances (PFASs). https://dtsc.ca.gov/scp/food-packaging-containing-pfass/

California State Legislature. (2022). Bill to phase out PFAS in textiles and apparel. https://www.leginfo.ca.gov/billtext/2022/PFAS_Textiles_Ban

ChemSec. (2023, December 5). Swedish court ruling unlocks potential for more PFAS lawsuits. https://chemsec.org/swedish-court-ruling-unlocks-potential-for-more-pfas-lawsuits/

CIRS Group. (2022, September 30). China to prohibit the import and export of PFOS products in 2024. https://www.cirs-group.com/en/chemicals/china-to-prohibit-the-import-and-export-of-pfos-products-in-2024

Corrs Chambers Westgarth. (2021, March 1). NSW the latest state to ban PFAS in firefighting foam.

https://www.corrs.com.au/insights/nsw-the-latest-state-to-ban-pfas-in-firefighting-foam

Cousins, I. T., Vestergren, R., Wang, Z., Scheringer, M., & McLachlan, M. S. (2019). The precautionary principle and chemicals management: The example of perfluoroalkyl acids in groundwater. Environmental Science & Technology, 53(5), 2301–2307. https://doi.org/10.1021/acs.est.8b06365

Cousins, I. T., DeWitt, J. C., Glüge, J., Goldenman, G., Herzke, D., Lohmann, R., Ng, C. A., Scheringer, M., & Wang, Z. (2019). The concept of essential use for determining when uses of PFAS can be phased out. Environmental Science: Processes & Impacts, 21(11), 1803-1815. https://doi.org/10.1039/c9em00163h

Cousins, I. T., Goldenman, G., Herzke, D., Lohmann, R., Miller, M., Ng, C. A., ... & Wang, Z. (2022). The concept of essential use for determining when uses of PFAS can be phased out. Environmental Science: Processes & Impacts, 24(1), 34–46. https://doi.org/10.1039/D1EM00391G

Dale, A., Siddique, S., Alvarez, J. M., & Mahoney, L. (2022). Wildfire-induced PFAS emissions: Atmospheric transport and deposition risks. Journal of Environmental Science & Technology, 56(4), 2258-2274. https://doi.org/10.1021/acs.est.1c08912

Danish Ministry of Environment and Food. (2020, May 28). Denmark moves ahead with PFAS ban in FCMs. https://foodpackagingforum.org/news/denmark-moves-ahead-with-pfas-ban-in-fcms

Domingo, J. L., Nadal, M., & Perelló, G. (2021). Human exposure to per- and polyfluoroalkyl substances (PFAS) through drinking water: A global review. Environmental Research, 195, 110866. https://doi.org/10.1016/j.envres.2021.110866

DuPont. (2023). Dark Waters: A film based on a true story. https://www.dupont.com/darkwaters.html

Ellison, G. (2021, May 7). 3M sues Michigan, seeks to invalidate PFAS drinking water rules. MLive. https://www.mlive.com/public-interest/2021/05/3m-sues-michigan-seeks-to-invalidate-pfas-drinking-water-rules.html

Enhesa. (2022, March 15). Tracking global PFAS regulations. https://www.enhesa.com/resources/article/global-pfas-regulation-managing-forever-chemicals/

Environmental Working Group. (2022). Mapping PFAS contamination. https://www.ewg.org

Environmental Working Group. (2022). PFAS contamination in U.S. states: An interactive map. https://www.ewg.org/pfasmap

Environmental Working Group. (2023). Major PFAS settlement leaves taxpayers on the hook for cleanup costs. https://www.ewg.org/news-insights/news/2023/06/pfas-settlement-still-leaves-taxpayers-hook

Envirotech Online. (2023, August 31). Is China moving towards stricter PFAS regulation? https://www.envirotech-online.com/news/pfas-analysis/105/international-environmental-technology/is-china-moving-towards-stricter-pfas-regulation/61076

European Chemicals Agency. (n.d.). Per- and polyfluoroalkyl substances (PFAS). Retrieved from https://echa.europa.eu/hot-topics/perfluoroalkyl-chemicals-pfas

European Chemicals Agency (ECHA). (2023, February 7). ECHA publishes PFAS restriction proposal. https://echa.europa.eu/-/echa-publishes-pfas-restriction-proposal

European Chemicals Agency. (2023). PFAS: Persistent and widespread concerns. https://echa.europa.eu/hot-topics/perfluoroalkyl-chemicals-pfas

European Chemicals Agency (ECHA). (n.d.). Per- and polyfluoroalkyl substances (PFAS). https://echa.europa.eu/hot-topics/perfluoroalkyl-chemicals-pfas

European Commission. (2024, September 15). Commission restricts use of a sub-group of PFAS chemicals. Retrieved from https://ec.europa.eu/commission/presscorner/detail/en/ip_24_4763

European Environmental Bureau. (n.d.). PFAS. Retrieved from https://eeb.org/work-areas/industry-health/pfas/

Evich, M. G., Davis, M. J. B., McCord, J. P., Acrey, B., Awkerman, J. A., Knappe, D. R. U., Lindstrom, A. B., Speth, T. F., Stevens, C. T., Strynar, M. J., Wang, Z., Weber, E. J., Henderson, W. M., & Washington, J. W. (2022). Per- and polyfluoroalkyl substances in the environment. *Science, 375*(6580), eabg9065. https://doi.org/10.1126/science.abg9065

Food & Water Watch. (2023). The multi-million-dollar effort to sway lawmakers on PFAS. https://www.foodandwaterwatch.org/2023/11/07/pfas-lobbying/

Food Packaging Forum. (2020). EU publishes PFOA regulation. Retrieved from https://foodpackagingforum.org/news/eu-publishes-pfoa-regulation

Gleiss Lutz. (2023). PFAS restriction proposal on the EU level. Retrieved from https://www.gleisslutz.com/en/news-events/know-how/pfas-restriction-proposal-eu-level

Grandjean, P., & Clapp, R. (2015). Perfluorinated alkyl substances: Emerging insights into health risks. New Solutions: A Journal of Environmental and Occupational Health Policy, 25(2), 147–163. https://doi.org/10.1177/1048291115590506

Grandjean, P., Timmermann, C. A. G., Kruse, M., Nielsen, F., Vinholt, P. J., Boding, L., ... & Heilmann, C. (2023). Immunotoxicity of perfluorinated alkylates: Epidemiological evidence for reduced antibody responses to vaccinations. Environmental Health Perspectives, 131(1), 117–123. https://doi.org/10.1289/EHP10557

Halaly, A. (2025, February 5). Cancer-causing 'forever chemicals' abound throughout the American West, study finds. Las Vegas Review-Journal. https://www.reviewjournal.com/local/local-nevada/cancer-causing-forever-chemicals-abound-throughout-the-american-west-study-finds-3274422/

Health Canada. (2023, August 9). Objective for Canadian drinking water quality per- and polyfluoroalkyl substances. https://www.canada.ca/en/health-canada/services/publications/healthy-living/objective-drinking-water-quality-per-polyfluoroalkyl-substances.html

Hu, X. C., Andrews, D. Q., Lindstrom, A. B., Bruton, T. A., Schaider, L. A., Grandjean, P., ... Sunderland, E. M. (2016). Detection of poly- and perfluoroalkyl substances (PFASs) in U.S. drinking water linked to industrial sites, military fire training areas, and wastewater treatment plants. Environmental Science & Technology Letters, 3(10), 344–350. https://doi.org/10.1021/acs.estlett.6b00260

International Chemical Regulatory and Law Review. (2023). Global regulations around PFAS: The past, the present, and the future. https://icrl.lexxion.eu/article/icrl/2023/1/4/display/html

Interstate Technology & Regulatory Council. (2023). PFAS fact sheets and regulatory framework. https://www.itrcweb.org/PFAS

Intertek. (2024, December 10). U.S. California AB 347 regulation of PFAS in consumer products.

https://www.intertek.com/products-retail/insight-bulletins/2024/u.s.-california-ab-347-regulation-of-pfas-in-consumer-products

iPoint-systems. (2023, March 10). PFAS Regulation in the EU: From POP to REACH. Retrieved from https://go.ipoint-systems.com/blog/pfas-eu-pop-reach

JD Supra. (2024). State level PFAS regulations are coming: Are you ready? https://www.jdsupra.com/legalnews/state-level-pfas-regulations-are-coming-4241786/

Johns Hopkins APL. (2023, May 18). Johns Hopkins APL explores alternatives to PFAS in firefighting foams. Retrieved from https://www.jhuapl.edu/news/news-releases/230518b-apl-explores-pfas-free-firefighting-foams

Kissa, E. (Ed.). (2001). Fluorinated surfactants and repellents (Vol. 97). CRC Press.

Kwiatkowski, C. F., Andrews, D. Q., Birnbaum, L. S., Bruton, T. A., DeWitt, J. C., Knappe, D. R. U., & Miller, M. F. (2020). Scientific basis for managing PFAS as a chemical class. Environmental Science & Technology Letters, 7(8), 532–543. https://doi.org/10.1021/acs.estlett.0c00255

Lang, J. R., Allred, B. M., Field, J. A., Levis, J. W., & Barlaz, M. A. (2017). National estimate of per- and polyfluoroalkyl substance (PFAS) release to U.S. municipal landfill leachate. Environmental Science & Technology, 51(4), 2197–2205. https://doi.org/10.1021/acs.est.6b05005

Lerner, S. (2018). DuPont and the chemistry of deception. The Intercept. https://theintercept.com/2018/11/20/dupont-chemours-pfoa-pfas-pollution/

Liu, Y., D'Agostino, L. A., Qu, G., Jiang, G., & Martin, J. W. (2021). High-resolution mass spectrometry (HRMS) methods for

nontarget discovery and characterization of PFAS. TrAC Trends
in Analytical Chemistry, 143, 116382.
https://doi.org/10.1016/j.trac.2021.116382

Liu, J., Mejia Avendaño, S., Buck, R. C., & Murphy, J. (2021). Global
occurrence, sources, and transport pathways of per- and
polyfluoroalkyl substances (PFAS) in the environment: A review.
Critical Reviews in Environmental Science and Technology,
51(5), 445–467.
https://doi.org/10.1080/10643389.2019.1635452

Mahler, A., Battaglin, W., Nakagaki, N., & Bradley, P. (2023). Costs
and considerations for addressing PFAS in U.S. public water
systems. Journal of Environmental Quality, 52(1), 20–36.
https://doi.org/10.2134/jeq2023.12

Michigan Department of Environment, Great Lakes, and Energy
(EGLE). (2020). PFAS contamination and fishing advisories in
Michigan. https://www.michigan.gov/egle

Michigan Department of Environment, Great Lakes, and Energy.
(2020, August 3). PFAS drinking water rules.
https://www.michigan.gov/egle/about/organization/drinking-
water-and-environmental-health/community-water-supply/pfas-
drinking-water-rules

Michigan PFAS Action Response Team. (n.d.). Maximum
contaminant levels (MCLs).
https://www.michigan.gov/pfasresponse/drinking-water/mcl

Maine Legislature. (2021). An Act To Stop Perfluoroalkyl and
Polyfluoroalkyl Substances Pollution (LD 1503).
https://www.mainelegislature.org/legis/bills/getPDF.asp?item=
5&paper=HP1113&snum=130

National Conference of State Legislatures. (2023, October 15). Per-
and polyfluoroalkyl substances (PFAS) | State legislation and

regulations. https://www.ncsl.org/environment-and-natural-resources/per-and-polyfluoroalkyl-substances

National Institute for Occupational Safety and Health. (2023). PFAS and worker health. Centers for Disease Control and Prevention. https://www.cdc.gov/niosh/pfas/about/index.html

New Jersey Department of Environmental Protection. (2020). PFAS standards and regulations. https://dep.nj.gov/pfas/standards/

New York State Department of Environmental Conservation. (2016). Per- and polyfluoroalkyl substances (PFAS). https://dec.ny.gov/environmental-protection/site-cleanup/pfas

New York State Department of Environmental Conservation. (2020). Per- and polyfluoroalkyl substances (PFAS) drinking water regulations. https://dec.ny.gov/environmental-protection/site-cleanup/pfas

Nguyen, V. T., Reinhard, M., & Gin, K. Y.-H. (2020). Fate and removal of poly- and perfluoroalkyl substances (PFASs) in water and wastewater treatment processes. Water Research, 182, 115982. https://doi.org/10.1016/j.watres.2020.115982

Oil Technics. (n.d.). International firefighting foam PFAS bans. https://www.firefightingfoam.com/news/international-pfas-bans/

Osler, Hoskin & Harcourt LLP. (2022, September 27). Regulation of 'forever chemicals' (PFAS) in Canada. https://www.osler.com/en/insights/updates/regulation-of-forever-chemicals-pfas-in-canada/

Peaslee, G. F., Wilkinson, J. T., McGuinness, S. R., Tighe, M., Caterisano, N., Lee, S., Gonzales, A., Roddy, M., Mills, S., & Mitchell, K. (2021). Fluorinated compounds in North American cosmetics. Environmental Science & Technology Letters, 8(7), 538–544. https://doi.org/10.1021/acs.estlett.1c00240

Pillsbury Winthrop Shaw Pittman LLP. (2023). Maine modifies its sweeping PFAS law. https://www.pillsburylaw.com/en/news-and-insights/maine-pfas-law.html

Pinsent Masons. (2023, December 12). Australia to introduce national controls for PFAS 'forever chemicals'. https://www.pinsentmasons.com/out-law/analysis/national-controls-for-pfas-forever-chemicals

Post, G. B., Cohn, P. D., & Cooper, K. R. (2012). Perfluorooctanoic acid (PFOA), an emerging drinking water contaminant: A critical review of recent literature. Environmental Research, 116, 93–117. https://doi.org/10.1016/j.envres.2012.03.007

Pulitzer Center. (2023, October 15). The forever lobbying project exposes the real cost of PFAS pollution. https://pulitzercenter.org/stories/forever-lobbying-project-exposes-real-cost-pfas-pollution-environment-science-and-politics

National Institute of Environmental Health Sciences (NIEHS). (2022). PFAS and agriculture: Assessing the economic and environmental impacts. https://www.niehs.nih.gov

Nguyen, V. T., Reinhard, M., & Gin, K. Y.-H. (2020). Fate and removal of poly- and perfluoroalkyl substances (PFASs) in water and wastewater treatment processes. Water Research, 182, 115982. https://doi.org/10.1016/j.watres.2020.115982

Post, G. B., Cohn, P. D., & Cooper, K. R. (2012). Perfluorooctanoic acid (PFOA), an emerging drinking water contaminant: A critical review of recent literature. Environmental Research, 116, 93–117. https://doi.org/10.1016/j.envres.2012.03.007

Powley, C. R., Michalczyk, M. J., Kaiser, M. A., & Buxton, L. W. (2005). Determination of perfluorooctanoate (PFOA) in water by ion-pair extraction and LC-MS-MS. Journal of Chromatography

A, 1073(1–2), 107–119.
https://doi.org/10.1016/j.chroma.2005.03.112

Reuters. (2024, April 15). New PFAS lawsuit cites EPA's 'forever chemicals' drinking water rules.
https://www.reuters.com/legal/litigation/new-pfas-lawsuit-cites-epas-forever-chemicals-drinking-water-rules-2024-04-15/

Rich, N. (2016). The lawyer who became DuPont's worst nightmare. The New York Times Magazine.
https://www.nytimes.com/2016/01/10/magazine/the-lawyer-who-became-duponts-worst-nightmare.html

Roland Berger. (2023, May 15). Understanding new PFAS regulations in the U.S. and EU.
https://www.rolandberger.com/en/Insights/Publications/Opportunities-challenges-in-new-regulations-of-forever-chemicals.html

Safer States. (2024, December 11). More than half of US state attorneys general have taken action against PFAS manufacturers and key users. https://www.saferstates.org/press-room/more-than-half-of-us-state-attorneys-general-have-taken-action-against-pfas-manufacturers

Schaider, L. A., et al. (2017). Fluorinated compounds in U.S. fast food packaging. Environmental Science & Technology Letters, 4(3), 105–111. https://doi.org/10.1021/acs.estlett.6b00435

Schultz, M. M., Higgins, C. P., Huset, C. A., Luthy, R. G., Barofsky, D. F., & Field, J. A. (2006). Fluorochemical mass flows in a municipal wastewater treatment facility. Environmental Science & Technology, 40(23), 7350–7357.
https://doi.org/10.1021/es061025m

Schwichtenberg, G., et al. (2023). Sociodemographic factors are associated with the abundance of PFAS in U.S. drinking water systems. Environmental Science & Technology.
https://pubs.acs.org/doi/abs/10.1021/acs.est.2c07255

Sharma, P., Patel, M., & Verma, S. (2023). Advances in electrochemical destruction of PFAS: Challenges and future prospects. Journal of Environmental Science and Technology, 58(3), 215–230. https://doi.org/10.1021/jes2023.09

Sims, D. B., Monk, J. R., Woldetsadik, D., Hudson, A. C., Buch, A. C., Garner, M. C., et al. (2025). Per- and polyfluoroalkyl substances (PFAS) in the rivers of the Western United States. International Journal of Environmental Science and Technology. https://doi.org/10.1007/s13762-024-06269-1

Singh, R., Kumar, N., & Dasgupta, S. (2023). Plasma-based PFAS remediation: Advances and economic considerations. Environmental Science & Technology Letters, 10(4), 451–459. https://doi.org/10.1021/acs.estlett.3c00218

Stockholm Convention. (n.d.). Overview - Stockholm Convention. Retrieved from https://chm.pops.int/Implementation/IndustrialPOPs/PFAS/Overview/tabid/5221/Default.aspx

Sunderland, E. M., Hu, X. C., Dassuncao, C., Tokranov, A. K., Wagner, C. C., & Allen, J. G. (2019). A review of the pathways of human exposure to poly- and perfluoroalkyl substances (PFASs) and significant policy challenges. Environmental Toxicology and Chemistry, 38(2), 241–257. https://doi.org/10.1002/etc.4373

Toxics Use Reduction Institute (TURI). (2025). Environmental Justice Through Toxics Use Reduction: Opportunities in Massachusetts. https://www.turi.org/wp-content/uploads/2025/01/TURI-2025-EJ-report-250122.pdf

Troutman Pepper. (2020, August 3). Michigan creates new PFAS in drinking water standards. https://www.troutman.com/insights/michigan-creates-new-pfas-in-drinking-water-standards.html

University of Maine News. (2025, January 17). PFAS-free firefighting biogel featured in Press Herald. Retrieved from https://umaine.edu/news/blog/2025/01/17/pfas-free-firefighting-biogel-featured-in-press-herald/

U.S. Environmental Protection Agency. (2022). Our current understanding of the human health and environmental risks of PFAS. https://www.epa.gov/pfas

U.S. Environmental Protection Agency. (2023a). EPA's actions to address PFAS. https://www.epa.gov/pfas

U.S. Environmental Protection Agency. (2023b). PFAS Strategic Roadmap: EPA's commitments to action 2021–2024. https://www.epa.gov/system/files/documents/2023-10/pfas-roadmap_final-508.pdf

U.S. Environmental Protection Agency. (2024a, April 26). Per- and polyfluoroalkyl substances (PFAS). https://www.epa.gov/pfas

U.S. Environmental Protection Agency. (2024b, April 10). Biden-Harris administration finalizes first-ever national drinking water standard to protect 100M people from PFAS pollution. https://www.epa.gov/newsreleases/biden-harris-administration-finalizes-first-ever-national-drinking-water-standard

U.S. Government Accountability Office. (2023). PFAS litigation and regulatory updates. https://www.gao.gov/reports/pfas-regulations

Wang, Z., Cousins, I. T., Scheringer, M., & Hungerbühler, K. (2017). Hazard assessment of fluorinated alternatives to long-chain perfluoroalkyl acids (PFAAs) and their precursors: Status quo, ongoing challenges, and possible solutions. Environment International, 75, 172–179. https://doi.org/10.1016/j.envint.2014.11.013

Wang, Z., DeWitt, J. C., Higgins, C. P., & Cousins, I. T. (2017). A never-ending story of per- and polyfluoroalkyl substances (PFASs)? Environmental Science & Technology, 51(5), 2508-2518. https://doi.org/10.1021/acs.est.6b04863

Wee, S. Y., & Aris, A. Z. (2023). Environmental impacts, exposure pathways, and health effects of PFOA and PFOS. Ecotoxicology and Environmental Safety, 267, 115663. https://doi.org/10.1016/j.ecoenv.2023.115663

Young, C. J., Mabury, S. A., & Martin, J. W. (2007). Atmospheric perfluorinated acid precursors: Chemistry, occurrence, and impacts. Environmental Science & Technology, 41(7), 2180–2186. https://doi.org/10.1021/es0617105

Appendix A

Glossary of PFAS-related terms

This glossary provides definitions for key terms related to per- and polyfluoroalkyl substances (PFAS), their environmental and health impacts, and regulatory frameworks. Understanding these terms is essential for grasping the full scope of the PFAS crisis and ongoing efforts to mitigate its effects.

Adsorption – The process by which PFAS chemicals adhere to the surface of a solid, such as activated carbon or ion exchange resins used in water treatment.

AFFF (Aqueous Film-Forming Foam) – A firefighting foam containing PFAS, widely used by the military, airports, and industrial facilities, and a major source of PFAS contamination in groundwater.

Analytical Detection Limit – The lowest concentration of a substance that can be reliably measured by laboratory instruments when testing for PFAS.

Bioaccumulation – The buildup of PFAS chemicals in living organisms over time, leading to increased concentrations in blood, organs, and tissues.

Biodegradation – The natural breakdown of substances by microorganisms. PFAS are highly resistant to biodegradation, contributing to their classification as "forever chemicals."

Carcinogenicity – The ability of a substance to cause cancer. Some PFAS compounds, such as PFOA and PFOS, have been linked to increased cancer risks in humans.

CERCLA (Comprehensive Environmental Response, Compensation, and Liability Act) – Also known as the Superfund law, CERCLA allows the U.S. government to clean up hazardous waste sites and hold polluters responsible for contamination. PFAS are being considered for designation as hazardous substances under this law.

Chemical Fingerprinting – A technique used to identify the specific PFAS compounds present in contaminated water, soil, or biological samples, helping to trace pollution sources.

Detection Limit – The smallest amount of PFAS that can be identified and quantified using laboratory analysis. Different methods have varying detection limits.

Drinking Water Advisory Level – A non-enforceable health guideline issued by agencies such as the EPA, indicating the maximum safe concentration of PFAS in drinking water.

EPA (Environmental Protection Agency) – The U.S. federal agency responsible for regulating environmental pollutants, including efforts to study and set limits on PFAS contamination.

Exposure Pathways – The routes through which humans and wildlife are exposed to PFAS, including drinking water, food, dust, air, and skin contact.

Fluoropolymer – A class of PFAS used in coatings, electronics, and industrial applications. Though chemically stable, fluoropolymers raise concerns due to their manufacturing and disposal impacts.

Food Contact Materials – Packaging, cookware, and processing equipment that may contain PFAS to provide non-stick or grease-resistant properties, potentially leading to ingestion.

Granular Activated Carbon (GAC) – A common water filtration technology that removes PFAS by adsorbing them onto porous carbon surfaces. GAC is effective but requires frequent replacement.

Groundwater Contamination – The presence of PFAS in underground water sources, often caused by industrial discharge, landfill leachate, and firefighting foam runoff.

Half-Life – The time it takes for half of a PFAS compound to be eliminated from the body or environment. Some PFAS have half-lives of years to decades, contributing to their persistence.

Hazardous Substance Designation – A classification under environmental laws that mandates stricter regulations and liability for PFAS contamination. Efforts are underway to list PFAS under CERCLA.

Ion Exchange Resins – A water treatment technology that removes PFAS by replacing them with less harmful ions. These resins can be engineered for high PFAS selectivity and regeneration.

Leachate – Contaminated liquid that drains from landfills, carrying PFAS and other pollutants into groundwater or surface water.

Limit of Quantification (LOQ) – The lowest level at which a laboratory can accurately measure the concentration of PFAS in a sample.

Mass Spectrometry (LC-MS/MS) – A highly sensitive laboratory technique used to detect and quantify PFAS compounds in environmental and biological samples.

Micropollutant – A contaminant present at trace levels in the environment but capable of causing long-term health and ecological effects, such as PFAS.

Non-Stick Coatings – Products such as Teflon™ cookware, where PFAS are used to provide heat-resistant, non-stick properties. Older coatings have been found to degrade over time, releasing PFAS.

National Primary Drinking Water Regulation (NPDWR) – The legally enforceable standard that the EPA is considering for regulating PFAS levels in public water supplies.

Oxidative Breakdown – A process where certain PFAS compounds transform under high-oxygen conditions. However, most PFAS are resistant to oxidation, making them difficult to degrade.

PFAS (Per- and Polyfluoroalkyl Substances) – A broad class of synthetic chemicals known for their water-, stain-, and grease-resistant properties, but notorious for their persistence and toxicity.

PFOA (Perfluorooctanoic Acid) – A long-chain PFAS formerly used in Teflon™, firefighting foams, and industrial coatings, linked to cancer, immune suppression, and reproductive harm.

PFOS (Perfluorooctane Sulfonic Acid) – Another legacy PFAS widely used in Scotchgard™, firefighting foams, and textiles, now phased out in many countries but still present in the environment.

Reverse Osmosis (RO) – One of the most effective water treatment technologies for PFAS removal, using a semi-permeable membrane to filter out even the smallest PFAS molecules.

Short-Chain PFAS – PFAS compounds with fewer carbon-fluorine bonds, marketed as "safer alternatives" but still highly persistent and mobile in water.

Stockholm Convention on Persistent Organic Pollutants – A global treaty aimed at banning and restricting hazardous chemicals, under which some PFAS are being considered for phase-out.

Toxicological Profile – A scientific evaluation of a chemical's health effects, such as PFAS toxicity on liver function, immune response, and cancer risk.

Total Organic Fluorine (TOF) – A broad measure of all fluorinated compounds in a sample, used to estimate PFAS contamination when specific compounds are unknown.

Wastewater Treatment Plant (WWTP) – Facilities that process sewage and industrial discharge. Many WWTPs are not designed to remove PFAS, leading to their spread in effluent and biosolids.

Water Solubility – The degree to which a substance dissolves in water. Some PFAS are highly soluble, making them more mobile in groundwater and drinking water supplies.

Appendix B

Key studies, reports, and databases

This appendix provides an overview of foundational studies, landmark reports, and critical databases that have shaped our understanding of PFAS contamination, health risks, regulatory responses, and remediation efforts. These resources offer scientists, policymakers, environmental advocates, and the public access to credible, data-driven insights into the extent and impact of PFAS pollution worldwide.

Key Studies on PFAS Contamination and Health Impacts

C8 Science Panel (2005–2013) – The Link Between PFAS and Human Health

One of the most significant PFAS health studies, the C8 Science Panel, investigated the effects of PFOA exposure in communities near DuPont's Washington Works plant in West Virginia. This extensive research, conducted over eight years, analyzed data from 69,000 residents and found probable links between PFOA exposure and six diseases, including kidney and testicular cancer, ulcerative colitis, thyroid disease, and high cholesterol. The findings played a key role in litigation against DuPont and set the stage for broader PFAS research and regulation.

📄 *Findings published in:* Environmental Health Perspectives (2013)

🔗 Website: www.c8sciencepanel.org

Grandjean & Clapp (2015) – Changing Interpretations of PFAS Health Risks

This study highlighted how regulatory agencies underestimated the toxicity of PFAS for decades. The authors reviewed historical data, corporate disclosures, and epidemiological findings, concluding that even low-dose PFAS exposure posed significant health risks. The research underscored the need for stricter regulations and a precautionary approach to chemical policy.

Published in: Public Health Reports, 130(4), 318-327

DOI: 10.1177/003335491513000411

Hu et al. (2016) – PFAS in U.S. Drinking Water

A groundbreaking study that analyzed nationwide PFAS contamination in drinking water. Researchers tested public water systems serving 66% of the U.S. population and found PFAS contamination exceeding EPA advisory limits in at least 6 million Americans' drinking water. The study accelerated state-led regulatory efforts and increased public awareness of the widespread nature of PFAS pollution.

Published in: Environmental Science & Technology Letters, 3(10), 344-350

DOI: 10.1021/acs.estlett.6b00260

Pelch et al. (2019) – Systematic Evidence Review on PFAS Toxicity

A comprehensive meta-analysis of over 740 studies, summarizing the toxic effects of PFAS on human health. This research confirmed PFAS exposure is associated with immune suppression, liver damage, developmental toxicity, and endocrine disruption, reinforcing calls for federal PFAS regulation.

259

📄   *Published   in:*   Environmental   Health,   18(1),   1-13

🔗 DOI: 10.1186/s12940-019-0484-0

Landmark Reports on PFAS Regulation and Environmental Impact

U.S. EPA PFAS Strategic Roadmap (2021–2024)

The EPA's most comprehensive PFAS action plan to date, outlining key regulatory steps, research priorities, and enforcement strategies to reduce PFAS exposure. The roadmap includes plans for enforceable drinking water limits, hazardous substance designations, and expanded   monitoring   requirements.

📄   *Issued   by:* U.S. Environmental Protection Agency (2021)

🔗 Website: www.epa.gov/pfas

ATSDR Toxicological Profile for PFAS (2018, Updated 2021)

The Agency for Toxic Substances and Disease Registry (ATSDR) published a detailed toxicological review of PFAS exposure pathways, health risks, and bioaccumulation patterns. The report established minimal risk levels (MRLs) for PFOA and PFOS, influencing state and   federal   guidelines   for   PFAS   exposure   limits.

📄 *Issued by:* ATSDR, U.S. Department of Health & Human Services

🔗 Website: www.atsdr.cdc.gov/toxprofiles/tp200.pdf

European Chemicals Agency (ECHA) PFAS Restriction Proposal (2023)

A groundbreaking proposal by five EU member states to ban thousands of PFAS chemicals across all industries under the EU REACH framework. This proposed restriction would be the largest

PFAS   ban   ever   attempted.
📄 *Issued by:* European Chemicals Agency (ECHA), 2023
🔗 Website: www.echa.europa.eu/pfas-restriction

Key Databases and Monitoring Resources

EPA PFAS Analytical Tools and Data Portal

A collection of publicly available databases, maps, and lab methods for detecting and tracking PFAS in the environment. Includes the PFAS Testing & Treatment Database, Drinking Water Monitoring Data, and PFAS Chemical Database.
🔗 Website: www.epa.gov/pfas-research

National PFAS Contamination Map (Environmental Working Group)

An interactive database tracking PFAS-contaminated sites across the U.S., including military bases, drinking water systems, and industrial discharge   sites.
🔗 Website: www.ewg.org/interactive-maps/pfas_contamination

ToxCast PFAS Screening Library

A database from the EPA's Toxicity Forecaster (ToxCast) program, which screens hundreds of PFAS chemicals for biological activity, potential toxicity, and endocrine disruption effects.
🔗 Website: www.epa.gov/chemical-research/toxcast

Stockholm Convention on Persistent Organic Pollutants (POPs) PFAS Inventory

A global registry of PFAS chemicals classified as persistent organic pollutants (POPs), documenting their environmental impacts, restrictions, and phase-out efforts.

🔗 Website: www.pops.int

This appendix serves as a starting point for deeper exploration into the science, policy, and advocacy efforts surrounding PFAS contamination. The studies, reports, and databases included here offer crucial insights into the extent of PFAS pollution, its health impacts, and global efforts to regulate and remediate these harmful chemicals. As research evolves and policies strengthen, these resources will continue to shape the fight against the forever chemical catastrophe.

Appendix C

Resources for advocacy and further reading

The fight against PFAS contamination requires informed advocacy, strong regulatory action, and continued scientific exploration. This appendix provides key resources for individuals, policymakers, scientists, and environmental organizations looking to educate themselves, support PFAS-free initiatives, and engage in meaningful advocacy.

Advocacy Organizations and Environmental Groups

Environmental Working Group (EWG)

The EWG has been a leading force in PFAS advocacy, conducting investigative research, tracking contamination hotspots, and pushing for stronger regulations. Their interactive PFAS contamination map is one of the most comprehensive tools for identifying impacted communities.

⌘ Website: www.ewg.org

Safer States

A coalition of state-level environmental groups working to ban PFAS in consumer products, drinking water, and industrial use. They provide resources for local policy action and legislative tracking.

🔗 Website: www.saferstates.org

Toxic-Free Future

A nonprofit advocating for PFAS-free alternatives in textiles, packaging, and household products. They provide consumer guides, corporate responsibility scorecards, and tools for petitioning major brands.

🔗 Website: www.toxicfreefuture.org

Green Science Policy Institute

A science-based advocacy group that works to bridge the gap between research and policy, providing scientific reports, educational webinars, and recommendations for PFAS phase-outs.

🔗 Website: www.greensciencepolicy.org

Earthjustice

A legal advocacy group that has led litigation efforts against PFAS polluters, including lawsuits against 3M, DuPont, and the Department of Defense for contamination cleanup and regulatory enforcement.

🔗 Website: www.earthjustice.org

Government Agencies and Regulatory Resources

U.S. Environmental Protection Agency (EPA) PFAS Hub

The EPA's central resource for PFAS regulations, health advisories, drinking water testing programs, and remediation efforts. The site includes policy updates, scientific research, and public health recommendations.

🔗 Website: www.epa.gov/pfas

Agency for Toxic Substances and Disease Registry (ATSDR) PFAS Information

A public health resource from the Centers for Disease Control and Prevention (CDC) providing PFAS exposure guidelines, toxicological profiles, and community health impact reports.

Website: www.atsdr.cdc.gov/pfas

European Chemicals Agency (ECHA) PFAS Restriction Proposal

The EU's proposal to ban thousands of PFAS chemicals, outlining scientific assessments, industry impact studies, and ongoing policy developments.

Website: www.echa.europa.eu/pfas-restriction

National Institute of Environmental Health Sciences (NIEHS) PFAS Research

A scientific hub featuring ongoing research into PFAS toxicity, bioaccumulation, and potential remediation technologies.

Website: www.niehs.nih.gov/health/topics/agents/pfas

PFAS-Free Consumer Guides

PFAS-Free Product List (Mamavation)

A continuously updated database of PFAS-free alternatives in cookware, clothing, cosmetics, food packaging, and more.

Website: www.mamavation.com/pfas-free-products

Silent Spring Institute PFAS in Consumer Goods Report

A science-based report on PFAS in everyday items, with recommendations for avoiding exposure.

🔗 Website: www.silentspring.org

PFAS-Free Fashion Directory

A consumer guide listing brands that have eliminated PFAS from their clothing, outdoor gear, and footwear.

🔗 Website: www.fashionforgood.com

Further Reading: Books and Reports

"Stain-Resistant, Water-Resistant, and Lethal: The Hidden Dangers of PFAS" – Callahan, S. (2023)

A detailed exposé on the history of PFAS, tracing corporate cover-ups, scientific discoveries, and the growing movement to ban these chemicals.

"Toxic: The Rot of American Science" – O'Connor, J. (2022)

An investigation into how industry-funded science has delayed chemical regulations and weakened public health protections.

"Forever Chemicals: Environmental Justice and the Fight Against PFAS Pollution" – Pelch, K. (2024)

A book exploring how marginalized communities have been disproportionately impacted by PFAS contamination and the legal battles for accountability.

How to Get Involved

Contacting Elected Officials

Legislators need to hear from their constituents about the urgency of PFAS regulation. Contacting local, state, and federal representatives through letters, emails, or town hall meetings can help push for stricter policies.

■▶ Find Your Representative: www.congress.gov/members

Signing Petitions and Supporting PFAS Bans

Several advocacy groups host ongoing petitions to ban PFAS in drinking water, food packaging, and consumer products.

Active Petitions:

- EWG's "Ban PFAS in Food Packaging" (www.ewg.org/petition)
- Toxic-Free Future's "PFAS-Free Fashion Now" (www.toxicfreefuture.org/pfas-free-petition)

Participating in Community Water Testing

Many local and state environmental organizations offer free or low-cost PFAS testing for drinking water.

Find Local PFAS Testing Programs: www.epa.gov/pfas

www.ingramcontent.com/pod-product-compliance
Lightning Source LLC
Chambersburg PA
CBHW062121020426
42335CB00013B/1050